FERMENTED FOODS AND BEVERAGES IN A GLOBAL AGE

BEER

FROM PRODUCTION TO DISTRIBUTION

FERMENTED FOODS AND BEVERAGES IN A GLOBAL AGE

Additional books and e-books in this series can be found on Nova's website under the Series tab.

FERMENTED FOODS AND BEVERAGES IN A GLOBAL AGE

BEER

FROM PRODUCTION TO DISTRIBUTION

ARMAND LEGAULT
EDITOR

Copyright © 2020 by Nova Science Publishers, Inc.

All rights reserved. No part of this book may be reproduced, stored in a retrieval system or transmitted in any form or by any means: electronic, electrostatic, magnetic, tape, mechanical photocopying, recording or otherwise without the written permission of the Publisher.

We have partnered with Copyright Clearance Center to make it easy for you to obtain permissions to reuse content from this publication. Simply navigate to this publication's page on Nova's website and locate the "Get Permission" button below the title description. This button is linked directly to the title's permission page on copyright.com. Alternatively, you can visit copyright.com and search by title, ISBN, or ISSN.

For further questions about using the service on copyright.com, please contact:
Copyright Clearance Center
Phone: +1-(978) 750-8400   Fax: +1-(978) 750-4470   E-mail: info@copyright.com.

NOTICE TO THE READER

The Publisher has taken reasonable care in the preparation of this book, but makes no expressed or implied warranty of any kind and assumes no responsibility for any errors or omissions. No liability is assumed for incidental or consequential damages in connection with or arising out of information contained in this book. The Publisher shall not be liable for any special, consequential, or exemplary damages resulting, in whole or in part, from the readers' use of, or reliance upon, this material. Any parts of this book based on government reports are so indicated and copyright is claimed for those parts to the extent applicable to compilations of such works.

Independent verification should be sought for any data, advice or recommendations contained in this book. In addition, no responsibility is assumed by the Publisher for any injury and/or damage to persons or property arising from any methods, products, instructions, ideas or otherwise contained in this publication.

This publication is designed to provide accurate and authoritative information with regard to the subject matter covered herein. It is sold with the clear understanding that the Publisher is not engaged in rendering legal or any other professional services. If legal or any other expert assistance is required, the services of a competent person should be sought. FROM A DECLARATION OF PARTICIPANTS JOINTLY ADOPTED BY A COMMITTEE OF THE AMERICAN BAR ASSOCIATION AND A COMMITTEE OF PUBLISHERS.

Additional color graphics may be available in the e-book version of this book.

Library of Congress Cataloging-in-Publication Data

ISBN: 978-1-53618-414-3

Published by Nova Science Publishers, Inc. † New York

Contents

Preface		vii
Chapter 1	Tailor-Made Concept for New Beer Types with High Biological Value *Vesela Shopska, Rositsa Denkova-Kostova, Kristina Ivanova and Georgi Kostov*	1
Chapter 2	The Revolution of Craft Beer *Cristina Calvo-Porral and Sergio Rivaroli*	77
Chapter 3	Effect of Processing on the Antioxidant Activity of Beer *Iuliana Aprodu*	123
Index		147

PREFACE

This compilation provides a critical review of a tailor-made concept for the production of beer with high biological value, reviewing the changes in phenolic capacity and antioxidant activity of malt and hops as the main phenolic component sources in beer. A conceptualization and characterization of craft beer and a full description of the "craft beer revolution" phenomenon is provided, elucidating the increasing consumption of this beverage. A concluding study assesses the antioxidant activity of beer mainly due to endogenous phenolic compounds with high bioavailability, Maillard reaction products and sulfites.

Chapter 1 - Food and beverages are building material and energy for people. They not only do satisfy hunger and thirst, but also supply the necessary micro- and macroelements, stimulants and regulators of various biological functions. In recent years, the so-called tailor-made (specially made for a particular purpose) concept entered food and beverage development resulting in the production of food and beverages intended for particular groups of consumers, providing not only nutritional functions but also additional health beneficial functions. This concept is applicable to different types of food and beverages and includes not only the choice of raw materials, but also the combination of technological processes for the production of food and beverages. Beer is one of the oldest beverages in the world. It is produced in many countries and consumed with enviable passion.

In recent years, its health benefits have been widely debated, and undoubtedly the consumption of beer in moderation, despite the presence of alcohol in it, has been proven healthy in a number of socially significant diseases. This is due to a number of its components - phenolic compounds, Maillard reaction products, etc., which have antioxidant effects. These components are derived from malt and hops, some of them change their bioavailability in the processes of malting, mashing and boiling, others undergo biochemical transformation during fermentation. The purpose of the present paper is to provide a critical overview of the tailor-made concept for the production of beer with high biological value. The paper reviews the changes in the phenolic capacity and the antioxidant activity of malt and hops as the main phenolic component sources in beer. Changes in phenolic content and antioxidant activity during malting, mashing, boiling and fermentation of beer wort are presented. Based on the data on the changes in the bioactivity during beer production, a tailor-made concept is presented. It is founded on the methods for mathematical and statistical processing and data on the biological value of beer and the changes that occur during technological processes, related to beer production. Moreover, as an element of brewing sustainability, data on possible applications of certain waste products from beer production to produce new types of tailor-made food, is also presented and discussed.

Chapter 2 - Craft beer has received much attention over the past two decades. The stability of consumption of beers and the rising interest towards craft beers is a clear signal that consumption patterns are changing. As a consequence, craft breweries have emerged as an alternative to industrial beer in many countries. The limited production and the use of unique non-traditional ingredients during the production process offer beers with original flavors, aromas, textures, and styles that are welcomed by consumers. Moreover, drinking craft beer is perceived as a sensory consumption experience and way to move away from industrial beer and to explore new taste experiences. So, the phenomenon of craft beer should be considered and analyzed. In this context, this chapter aims to describe the so-called "craft beer revolution," shedding some light on some of the most significant drivers of craft beer consumption that are changing the status quo of the alcoholic beverage sector. More precisely, this study aims to provide a conceptualization and characterization of craft beer and a full description of the "craft beer revolution" phenomenon, explaining the increasing consumption of this beverage. Besides, the chapter will analyze the craft beer production and distribution systems, characterized by its limited production, the use of non-traditional and unique ingredients, and a specific trade-off consumption. Finally, the chapter will examine the main drivers and motivations underlying craft beer consumption.

Chapter 3 - Beer is rich in bioactive compounds, mainly antioxidants, which provide health related benefits in case of moderate consumption. The antioxidant activity of beer is mainly due to the endogenous phenolic compounds with high bioavailability, Maillard reaction products and sulfites. The phenolics coming from malt account for up to 80% of the total phenolic content of beer and include phenolic acids and derivatives, anthocyanins, proanthocyanidins, lignans and lignin-related compounds. Additional phenolic acids, chalcones, flavonoids, catechins, and proanthocyanidins are derived from hops. In addition to the contribution to the formation of beer color, aroma and flavor, the final products of the Maillard reaction, melanoidins, exhibit strong antioxidant properties and other biological effects. Finally, the sulfites found in beer are fermentation by-products produced by yeasts while synthesizing of sulfur-containing amino acids. The amount of endogenous antioxidants varies with the beer type, being affected by genetic and agricultural factors in the raw material and by processing during malting and brewing. The technological steps influence in different ways the antioxidant activity of beer. The amount of phenolic compounds and the overall antioxidant activity are significantly affected during malting, mashing, fermentation and storage. Better antioxidant activity of beer and improved retention of the biologically active compounds can be obtained by selecting the appropriate raw materials and optimizing the brewing parameters.

In: Beer: From Production to Distribution ISBN: 978-1-53618-414-3
Editor: Armand Legault © 2020 Nova Science Publishers, Inc.

Chapter 1

TAILOR-MADE CONCEPT FOR NEW BEER TYPES WITH HIGH BIOLOGICAL VALUE

Vesela Shopska[1], Rositsa Denkova-Kostova[2], Kristina Ivanova[3] and Georgi Kostov[1],*

[1]Department "Wine and Beer Technology,"
University of Food Technologies,
Plovdiv, Bulgaria
[2]Department "Biochemistry and Molecular Biology,"
University of Food Technologies,
Plovdiv, Bulgaria
[3]Institute of Canning and Food Quality,
Agricultural Academy,
Plovdiv, Bulgaria

---

* Corresponding Author's Email: george_kostov2@abv.bg.

ABSTRACT

Food and beverages are building material and energy for people. They not only do satisfy hunger and thirst, but also supply the necessary micro- and macroelements, stimulants and regulators of various biological functions. In recent years, the so-called tailor-made (specially made for a particular purpose) concept entered food and beverage development resulting in the production of food and beverages intended for particular groups of consumers, providing not only nutritional functions but also additional health beneficial functions. This concept is applicable to different types of food and beverages and includes not only the choice of raw materials, but also the combination of technological processes for the production of food and beverages.

Beer is one of the oldest beverages in the world. It is produced in many countries and consumed with enviable passion. In recent years, its health benefits have been widely debated, and undoubtedly the consumption of beer in moderation, despite the presence of alcohol in it, has been proven healthy in a number of socially significant diseases. This is due to a number of its components - phenolic compounds, Maillard reaction products, etc., which have antioxidant effects. These components are derived from malt and hops, some of them change their bioavailability in the processes of malting, mashing and boiling, others undergo biochemical transformation during fermentation.

The purpose of the present paper is to provide a critical overview of the tailor-made concept for the production of beer with high biological value. The paper reviews the changes in the phenolic capacity and the antioxidant activity of malt and hops as the main phenolic component sources in beer. Changes in phenolic content and antioxidant activity during malting, mashing, boiling and fermentation of beer wort are presented.

Based on the data on the changes in the bioactivity during beer production, a tailor-made concept is presented. It is founded on the methods for mathematical and statistical processing and data on the biological value of beer and the changes that occur during technological processes, related to beer production.

Moreover, as an element of brewing sustainability, data on possible applications of certain waste products from beer production to produce new types of tailor-made food, is also presented and discussed.

Keywords: beer, tailor-made concept, bioactivity compounds

1. BASIC RAW MATERIALS AND PROCESSES IN THE PRODUCTION OF MALT AND BEER

1.1. Basic Raw Materials for the Production of Malt and Beer

1.1.1. Grain Raw Materials

The brewing industry is relatively conservative in terms of the raw materials used to produce the beverage. This is determined by the historical preconditions of beer production and the number of local and international legislative restrictions introduced by them. At the present stage of development, the main raw materials for beer production remain barley and wheat malt and their special varieties, offered for the production of beverages of different colors and flavors. For the brewing industry, the content of carbohydrates in the raw material is of major importance, although the taste, aroma and color of the beer are also decisive in the choice of raw material (Kabzev and Ignatov, 2011; Meussdoerffer and Zarnkow, 2009).

Barley (*Hordeum vulgare*) and barley malt meet best the requirements of the brewing industry. They contain high amount of carbohydrates, as well as all the enzyme systems necessary for beer production. In addition, barley structural parts prove to be a good lautering agent. For this reason, the use of barley as a raw material for beer production dates back to ancient times and has been established over the years. The introduction of the "Reinheitsgebot" accelerates the process of selection of barley with brewing qualities and establishes barley malt as the main raw material for beer production. The optimization of malt production processes leads to the production of raw material with optimal physicochemical composition, suitable for beer production. Malted raw materials are the main material for beer production, so it is good to make predictions about the suitability of different raw materials for malting (Kabzev and Ignatov, 2011; Meussdoerffer and Zarnkow, 2009).

Although the current Beer Purity Act, which stems from the original Reinheitsgebot, stipulates that bottom-fermented beers are made only from barley malt, and rice and corn are not defined as non-grain raw materials, wheat, rice, corn, sorghum and other raw materials are also used for beer production worldwide, including in Germany. This predetermines the production of different beer types (Kabzev and Ignatov, 2011; Meussdoerffer and Zarnkow, 2009).

The choice of grain raw material for beer production is determined by three main requirements: the physicochemical profile of the raw material, the possibility of obtaining the raw material in the area of beverage production and, last but not least, its price. And since the present chapter focuses on methods for optimizing and producing new beer assortments, we will note only the more important physicochemical aspects of selection. In the first place in terms of requirements is the starch content in the raw material, as well as the protein content. An important requirement is that the protein content does not exceed 9-12%, although there are deviations from this rule depending on the ratio of malted and unmalted raw materials used. The possibility of fine or coarse grinding of the raw material also determines its use. Therefore, wheat, rice and corn are mainly used in beer production along with barley malt. They meet best the three requirements and allow for the production of a wide range of assortments (Kabzev and Ignatov, 2011; Meussdoerffer and Zarnkow, 2009).

The use of raw materials other than barley malt requires enzyme preparation application to some extent. It should be noted that the enzymes contained in the malt are complementary to the substrate, i.e., they are adapted to the carbohydrates and proteins of the respective raw material, while the enzyme preparations containing proteases and glucohydrolases require purified form of starch and proteins. Since adjunct grists contain mainly unmodified endosperm, therefore contain dense granules of starch, thus, enzyme action is limited (Kabzev and Ignatov, 2011; Meussdoerffer and Zarnkow, 2009).

The main grain raw materials identified by the FAO are wheat, rice, barley, rye, millet, oats, sorghum, buckwheat, triticale, fonio. These raw materials are the main ones that are used in the brewing industry around the world. The choice is made based on the content of carbohydrates, proteins, fats, mashing method and other technological criteria (Kabzev and Ignatov, 2011; Meussdoerffer and Zarnkow, 2009).

The first main raw material to consider is barley and barley malt. Mainly two-row varieties of barley, which undergoes malting, are used for brewing. Unmalted barley is applied in an amount of up to 20%, which rather does not require the use of external enzymes. The unmalted barley is finely ground, which provides higher yield of extract. It is preferable to use barley varieties with low starch gelatinization temperature, as they do not require the addition of enzymes. Industrial enzymes are used when increasing the amount of unmalted barley, and one of the important conditions is to apply preparations with β-glucanase activity to reduce mash lautering problems. Alternative methods are extrusion or infrared light treatment. Unmalted barley can be roasted, which is typical of some Irish beer types. In addition, roasted barley and its extract comply with the Beer Purity Act and can be used as a colorant for beer production (Kabzev and Ignatov, 2011; Meussdoerffer and Zarnkow, 2009).

Another grain raw material used in the brewing industry is oats (*Avena sativa* L). In the middle Ages it was the main raw material for beer production, but it quickly lost its importance with the advent of barley. Due to the peculiarities of their structure, oat grains adsorb water quickly, which reduces the steeping time in oat malt production. Malting is carried out under similar conditions to barley malting. However, high losses of extract are observed as the flakes contain more sugars. Oats are low in fat, protein and β-glucans. Oats are characterized by low hydrolase activity, which is the main reason for their poor application in the brewing industry. However, some varieties are preferred in the brewing industry due to the low β-glucans content. The resulting oat wort has similar characteristics to that obtained from barley malt. The main advantage is the fast lautering process due to the large oat husks. The wort contains significant amounts of zinc and tryptophan. The resulting beer has good foaming capacity and stable turbidity. Oats are used mainly in the production of top-fermented beer types (Kabzev and Ignatov, 2011; Meussdoerffer and Zarnkow, 2009).

Millet and corn are botanically related. Millet is grown mainly in Africa, with only some varieties being used in the brewing industry. Pearl millet

(*Pennisetum glaucum* (L.) R. Br.) is used as a malted raw material for some cloudy beer types. The malting process varies from 1 to 5 days, at relatively high temperatures of 25-30°C. The malt has a relatively high β-amylase activity. Spring millet (*Setaria italica* (L.) P. Beauv.) has a relatively long germination period, which results in inhomogeneous malt with brewing characteristics close to those of other millet species. Fonio (*Digitaria exiliz*) is a millet type with small grains. It is used for cloudy beer types produced in Africa. It is used as unmalted raw material or combined with other millet malt types. Germination occurs at 30°C for 4 days as a result of which maximum enzymatic activity is achieved. Common millet (*Panicum miliaceum* L.) is a well-known brewing raw material. Germination proceeds quickly due to the small grains, but is slowed down by the steeping process, as the husk of the grain is relatively thick. The best characteristics are obtained at 27°C for 7 days. The resulting malt has relatively good brewing characteristics. Millet gives the malt mash a pleasant sweet aroma, which, however, is unstable and is lost in subsequent operations. When millet is used during mashing, it gives the beer a pleasant sweet aroma, which is not retained further in beer production. However, beer obtained with that type of millet or malt differs significantly from standard beer types (Kabzev and Ignatov, 2011; Meussdoerffer and Zarnkow, 2009).

Corn and its starch are suitable substitutes for malt, but they require a certain degree of purification before use. The most important part is the removal of the corn germ, as it contains a significant amount of fats, which can affect the fermentation process and the beer quality. Corn is applied in the form of semolina, the germ being removed during grinding. The saccharification and gelatinization of the starch takes place during mashing, but it is also possible to use pre-modified starch. Replacing up to 5% of malt with corn does not change the wort characteristics. Larger amounts lead to a change in the wort protein composition, which requires that the free aminonitrogen does not fall below 150 mg/dm^3. In case of reduced nitrogen content, the addition of nitrogen components during fermentation is recommended. A combined process with the addition of 10% malt raw material mixture to the adjunct mash and, respectively, mashing and liquefaction of the corn starch at 78°C is used for corn mashing. Corn could be malted, but the losses from germination are large, and the resulting raw material has low diastatic power (Kabzev and Ignatov, 2011; Meussdoerffer and Zarnkow, 2009).

Along with corn and barley, rice is the raw material most commonly used to replace barley malt in beer production. It is a starch-rich raw material - polished rice contains about 80-84% starch, 5-8% protein; 0.5-1% fat and 0.8-1.2% minerals. The rice should be washed several times before use in order to reduce fat. As a result of this reduction, some metabolites such as γ-nonalactone and 1-hexanol, which are specific components and are strongly felt in fermented beer, are also reduced. Rice varieties rich in amylopectin are suitable for beer production. Mashing methods which include boiling of the surrogate mash at 90-100°C or obtaining combined enzyme preparations are recommended due to the high gelatinization temperature of rice starch. On the other hand, low mashing temperatures can be applied to rice varieties with low gelatinization temperature, the regimes being close to those of barley mashing. Rice can also be malted, but steeping is relatively long. In practice, different varieties of rice are used. Characteristic of some of them, such as black rice, is that they lead to wort coloring due to the extraction of antocyans in it (Kabzev and Ignatov, 2011; Meussdoerffer and Zarnkow, 2009).

Rye is a raw material that is used primarily in its malted version. The resulting malt mash has an increased viscosity due to the presence of pentosans in it. For this reason, there are difficulties with the filtration process, as well as oxidation of the malt mash. Another problem associated with the increased viscosity is the difficulty of obtaining a crystal-colored beer. Beer with rye has a pleasant aroma, it is cloudy and is obtained only by top fermentation. Roasted rye malt is also used for beer production. It gives good coloring, has color in the range of 500-800 EBC units, close to the color of roasted spelt or roasted barley malt (Kabzev and Ignatov, 2011; Meussdoerffer and Zarnkow, 2009).

Sorghum is used for the production of lager and cloudy beers in Africa. It is applied both as unmalted and as malted raw material. The resulting sorghum malt has similar brewing characteristics to those of its barley analogue, but has a high gelatinization temperature and low diastatic power, due to the reduced β-amylase activity. Another characteristic feature is the low protein content in the wort. Unmalted sorghum is the third largest substitute for barley malt after corn and rice. The sorghum is ground in dry mills, and the germs are removed during the grinding. Beer wort obtained with unmalted sorghum has light color (Kabzev and Ignatov, 2011; Meussdoerffer and Zarnkow, 2009).

The production of wheat beer types is associated with the use of malted and unmalted wheat representatives (*Triticum* L.). Different representatives of this grain raw material are used - spelt, einkorn, emmer, kamut, triticale and others. The distribution of different wheat types in the world determines the different types of beer and beverages that are produced in the different parts of the world. For example, spelt is used in beer production in some parts of southern Germany (Swabian beer). Spelt malt has similar characteristics to that obtained from common wheat, so it can replace up to 70% of barley malt in the mixture. Roasted spelt malt has a coloration intensity of about 450-650 EBC units, which leads to less colored dark beer types compared to beer types produced with roasted barley malts. The beer obtained with spelt without husks has weak, smooth aroma. In recent years, the production of einkorn beer (*Triticum monococcum* L. - single-row wheat, with diploid grain with husk) is one of the new directions in the beer production industry. Storage, processing and malting of einkorn are difficult tasks due to the sensitive embryo. The protein content as well as the aminonitrogen are below average, although no such deviations are observed for the unmalted raw material. Another important feature is the low fermentation rate with respect to the extract. The wort obtained with the participation of einkorn, as well as the subsequent beer show good foam stability and a pleasant carbonated taste due to the good gas retention in the beverage. Emmer (*Triticum dicoccum Sch ü bl.*) and kamut (*Triticum durum* L., *Triticum turgidum* L. *and Triticum polonicum* L.) are two other representatives of wheat. They allow the production of different wheat beer types than the standard ones. Emmery malt gives high yield of extract and good time for saccharification, but poor fermentation ability. The protein content is within normal limits. The wort is without any peculiarities. The beer has a slightly sour taste, but when using top fermentation it differs from other similar beer types. Kamut is characterized by various problems - low yield of extract, problems with color and aminonitrogen content, despite the high protein content before malting. The diastatic power of malt is relatively high. The beer obtained with such malt has satisfactory sensory characteristics (Kabzev and Ignatov, 2011; Meussdoerffer and Zarnkow, 2009).

Triticale (*T. riticosecale* Wittmack) is a hybrid wheat type that has recently been used in the brewing industry. The triticale qualities for malting are close to those of rye. Color intensity and viscosity are not high, malt has a high diastatic power, but the α-amylase activity is below average. Unlike rye malt, triticale malt leads to constant beer turbidity. In addition, the qualities of top-fermented beers are satisfactory (Kabzev and Ignatov, 2011; Meussdoerffer and Zarnkow, 2009).

Classic wheat (*Triticum aestivum* L.) is used for the production of wheat beer types, both as malted and unmalted raw material. In Germany, malted wheat is used for beer production, while in Belgium, unmalted wheat is usually used for beer production, and it can replace up to 50% of the malt in beer. Belgian wheat beers usually ferment spontaneously. Since wheat has a low gelatinization temperature, malt enzymes are able to hydrolyze starch without boiling the raw material. Even standard bread flour can be used for beer production (*Kabzev and Ignatov, 2011; Meussdoerffer and Zarnkow, 2009*).

1.1.2. Hops and Hop Preparations

Hops (*Humulus lupulus* L.) are the second main raw material for beer production. Its specific components give the beverage pleasant bitterness and aroma, improve its foaming and foam-retention properties and increase its biological value and stability. Hops are from the hemp plant family (*Connabinaceae*), with different male and female inflorescences that grow on different plants. Female inflorescences, better known as hop cone flour, are used in brewing (Kabzev and Ignatov, 2011).

The specific ingredients of hops - hop bitterness and essential oils are used in beer production. They are responsible for the beer bitter taste and aroma. Some of these ingredients are extracted into the beer during the brewing process and in most cases they have significant impact on the course of the technological process. The accumulation of bitter substances in hops begins during its growing season. Initially, the lupolin glands accumulate mainly β-acids, some of which turn into α-form when the hops ripen. The ratio of α- and β-acids determines the differences between bitter and aromatic hop varieties. As a result of a number of studies, it has been proven that the content of α-acids is a varietal mark (Kabzev and Ignatov, 2011).

α-acids are found in trace amounts in all beer types. The main bitterness is determined by iso-α-acids, the isomerization process occurs at high temperatures during wort brewing. It is preceded by extraction of humolones (the main representatives of this group of acids). Compounds with higher solubility, stronger bitterness and higher resistance to oxidation are obtained as a result of isomerization (Kabzev and Ignatov, 2011).

Hops polyphenols are essential for the technological process. The content of polyphenols depends on the hops variety and the agro-climatic conditions of its cultivation. Hops contain hydroxybenzoic acid, hydroxycinnamic acid, proanthocyanidins, flavanols, quartcetin, flavonols and others. It contains gallic acid, p-hydrosibenzoic acid, vanilla acid, caffeic acid and others in different amounts, depending on its variety. Some of these components are responsible for the formation of beer taste and aroma, while others are related to the colloidal and biological stability of the beverage. In general, it has been found that the antioxidant potential is due to low molecular mass fractions (Kabzev and Ignatov, 2011).

1.1.3. Water

Water makes up 90% of the beverage. Water quality characteristics, determined by its physicochemical composition, are related to the final quality of the beer. Water composition plays fundamental role in the extraction of the components of hops and malt, determines the rate of enzymatic reactions, contributes to the beer buffering capacity and directly affects the taste and aroma of the finished beverage. There are different requirements for water, and to a large extent they are standardized for different beer types (Kabzev and Ignatov, 2011).

1.1.4. Brewer's Yeast

Brewer's yeast is responsible for the processes of alcohol fermentation and beer maturation. Beer taste and aroma profile are formed during these processes. Brewer's yeast belongs to: kingdom *Fungi*, type *Eumycota*, subtype *Ascomycotina*, class *Hemiascomycetes*, order *Endomycetales*, family *Saccharomycetaceae*, genus *Saccharomyces* (Boekhout and Pfaff, 2003). Depending on their behavior during fermentation, yeasts are divided into top- and bottom-fermenting, and according to the latest classifications, bottom-fermenting yeasts are part of the species *Saccharomyces pastorianus*, but in practice the name *Saccharomyces carlsbergensis* has been retained, while top-fermenting strains belong to the species *Saccharomyces cerevisiae* (Tenge, 2009; Shopska et al., 2016; Boekhout and Pfaff, 2003).

Brewer's yeast cells are round, oval or elongated, and their size varies from 2.5 to 21 μm (on different axes of the cell) depending on the yeast type and the cultivation conditions. Cells proliferate by budding, and by spore formation under adverse conditions (Shopska et al., 2016; Boulton and Quain, 2001; Briggs et al., 2004; Tenge, 2009).

Brewer's yeasts are heterotrophic, facultative anaerobes, absorbing carbohydrates (glucose, fructose, maltose, maltotriose, galactose, and raffinose), organic acids (lactic acid and succinic acid) and alcohols (they metabolize ethanol and glycerol in the presence of oxygen). They assimilate inorganic (ammonium) and organic (amino acids, peptides, amines, nucleotides) forms of nitrogen during their growth. The wort supplies the yeast with all the necessary minerals, except for Zn^{2+}. Therefore, Zn^{2+} often have to be added before fermentation. With regard to growth factors, it has been found that all brewer's yeasts need biotin and pantothenic acid for their growth (Shopska et al., 2016; Boulton and Quain, 2001; Briggs et al., 2004).

The two yeast types have different morphological and physiological characteristics. In budding, the top-fermenting yeasts form short chains, while in the bottom-fermenting ones the division of the two cells occurs immediately after the end of budding (Shopska et al., 2016; Tenge, 2009; Kunze, 2004). Perhaps the most significant difference is the ability to metabolize wort raffinose. Top-fermenting yeasts absorb only 1/3 of this carbohydrate (fructose), due to the lack of the enzyme α-D-galactosidase in their enzyme set, they cannot break down melibiose to glucose and galactose (Tenge, 2009). Bottom-fermenting yeasts fully assimilate raffinose and show higher activity in the utilization of galactose and maltotriose than top-fermenting yeasts (Briggs et al., 2004; Boulton and Quain, 2001; Shopska et al., 2016). The other main physiological difference is the fermentation temperature. *S. carlsbergensis* ferments best in the range of 7-15°C, while *S. cerevisiae* - at 18-25°C. The two types of brewer's yeast also differ in their flocculation ability and metabolism (Briggs et al., 2004; Boulton and Quain, 2001; Shopska et al., 2016; Tenge, 2009).

There are various requirements to yeasts, such as reproductive capacity, fermentation activity, flocculation capacity and appropriate metabolism. For beer production, strains with lower reproductive capacity, higher fermentation activity and appropriate metabolism are recommended. The high reproductive capacity leads to accelerated substrate consumption, increased biomass synthesis, production of more secondary metabolites and increased losses of hop bitter substances. As a result, beer of poor quality is obtained. Low reproductive capacity leads to a prolonged fermentation process, which increases the possibility of growth of contaminating microflora. The yeast reproductive ability depends on the wort composition, the oxygen content in the medium, the inoculum size, the pH, the fermentation temperature, the yeast metabolites (Shopska et al., 2016; Munroe, 2006). The fermentation activity depends on the genetic characteristics of the yeast strain, the yeast physiological state, the yeast cell surface, the nitrogen content in the cells and the environment, the yeast reproductive and flocculation ability (Shopska et al., 2016). Flocculation affects the rate and degree of fermentation, as well as the beer clarification. It depends on the yeast strain and the amount of yeasts, the wort composition, the fermentation temperature and aeration (Shopska et al., 2016; Briggs et al., 2004; Tenge, 2009). Brewer's yeast assimilates various substances from wort. Some of them are used for the synthesis of biomass, while others are used for the synthesis of various products that are essential for the quality of the resulting beer (Shopska et al., 2016).

1.2. Main Processes in Malt Production

Malt production takes place in several main stages. The first of them includes the delivery, qualification and primary cleaning of the raw material before its storage. Storage should take place after batching by variety and protein content. Primary cleaning is envisaged to remove mechanical damages that may have a negative impact during the storage process (Narziss, 1999; Briggs et al., 2004; Kunze, 2004). The storage of barley as the main raw material for malt production is carried out under controlled conditions (humidity up to 14% and temperature up to 12 °C). If necessary, drying is carried out before storage to reduce the moisture content (Kreisz, 2009).

Barley is stored in silos until its use. The quality characteristics are determined and the optimal steeping, germination and kilning modes are selected for each batch before its passing to the next technological operation. The barley is then sent to the cleaning plant by transport systems, which must ensure quality and non-destructive transport of the grains (Kreisz, 2009).

The homogeneity of the grain mass is important for malt quality. This requires the grains to go through a sorting stage. Barley is divided into three main qualities, in terms of grain width - first quality (over 2.5 mm), second quality (2.2-2.5 mm) and third quality (less than 2.2 mm). An element of this process is the secondary cleaning of the raw material and its cleaning from other crops, dust and other mechanical impurities. Malt is produced mainly from first quality barley, although it is technologically possible to use other raw materials such as wheat, corn, rice and others (Kreisz, 2009).

Increasing the water content in the grain is the main prerequisite for its germination (Kreisz, 2009). Germination takes place at a final humidity of steeped barley of 44-50%, and this humidity is called "degree of steeping." The steeping process and the ensuring of an optimal degree of steeping is essential for the enzyme catalytic action during germination. Although the increase in the degree of steeping increases enzyme biosynthesis in the grains, high humidity disrupts the grain husk and its permeability, which leads to embryo death (Kreisz, 2009).

The germination process begins at a water content of more than 30%, and its rate begins to be controlled during the steeping stage. The steeping process is controlled by temperature, steeping time, oxygen supply and grain retention periods in water and air (Kreisz, 2009).

Potable water in an amount of 2-3.5 m^3/t of raw material and temperature of 12-18°C is used for steeping. High water temperatures and long water breaks increase the process rate, but there is a risk of overwetting and the growth of microorganisms. In addition, the uptake of water from different parts of the grain is different. It has been found that the regions around the germ and the husk absorb more water than the endosperm. An important element of the steeping process is the oxygen supply to the grains. Compressed air is most often used for this purpose, as it simultaneously supplies the grains with oxygen and mixes the grain mass. Lack of oxygen provokes the accumulation of CO_2 in the grain and the poisoning of the grain. This process should be avoided, except in the production of certain types of acid malts, where this is desirable. Water uptake and subsequent germination strongly depend on the barley quality. Barley may show high germination capacity, but the energy might still not be high during steeping. This phenomenon is called latency and protects the grains from pre-germination without providing optimal conditions for this process (Kreisz, 2009).

The main stage of malt production is the germination. As a result, the following processes take place - hydrolysis of cell walls and proteins, accumulation of hydrolytic enzymes, hydrolysis of proteins to free aminonitrogen, reduction of potential losses of extract during respiration and grain growth. Quality green malt is obtained as a result of all these changes, and it is subjected to kilning to stabilize its properties. These goals are achieved by controlling the process time, the optimal degree of steeping, the process temperature, the ratio of oxygen and carbon dioxide content and the addition of gibberellic acid (Narziss, 1999; Briggs et al., 2004; Kunze, 2004; Kreisz, 2009).

The germination process is associated with the formation of roots of the grain and structural changes occurring in the main parts of the grain. Growth is heterogeneous and the grain undergoes significant physiological and physicochemical changes. As a result, products of decomposition of the endosperm and the aleurone layer accumulate - sugars, amino acids, metal ions, phosphates and others. The technological goals of these processes are control of protein hydrolysis, cytolysis and starch production. The accumulation of these compounds is important for the brewing process in beer production. For this reason, the right combination of temperature, humidity and germination time is important for the production technology (Narziss, 1999; Briggs et al., 2004; Kunze, 2004; Kreisz, 2009). The germination process and the changes that occur in the grains depend largely on the barley variety. It can be summarized that the following reactions take place in malt (Kreisz, 2009):

- The accumulation of enzymes starts after the first day of steeping and continues until the beginning of the kilning process. Different groups of enzymes accumulate at different rates during the germination process.
- The maximum amount of enzymes is reached on the sixth day after germination, and the amount of soluble nitrogen reaches a maximum at the beginning of the kilning process.
- The high water content leads to a higher degree of modification and to a higher process rate, but the final yield of extract is lower.
- Germination energy is higher at higher water content.
- Low germination temperatures favor the synthesis of enzymes and soluble nitrogen, while high temperatures accelerate respiration and lead to higher extract losses. This requires the use of variable temperatures in the different phases of germination.
- The addition of gibberellic acid stimulates the modifications in the grain, increases the enzyme activity and improves malt cytolysis. At the same time, the color intensity increases, but there are possible risks of uneven germination.

Enzymes are responsible for the hydrolysis of the cell walls, the proteins and the starch. The synthesis of α- and β-amylase and the breakdown of proteins and β-glucans are essential in the germination process. The breakdown of β-glucans in cell walls performed by β-glucan solubilase and endo-β-glucanase is very important to achieve rapid mash lautering. Since this is an enzymatic step, the breakdown of β-glucans into smaller water-soluble molecules, such as glucose, is favored by high moisture content and a temperature of about 19°C (Briggs et al., 1981; Briggs et al., 1982; Briggs et al., 2004).

The last main stage responsible for malt quality is the kilning of green malt. At this stage the product forms its organoleptic and physicochemical properties. The kilning process has the following main technological goals (Narziss, 1999; Briggs et al., 2004; Kunze, 2004; Kreisz, 2009):

- Ensuring long-term storage of the finished product by reducing the humidity from 42-48% to 3-3.5%.
- Termination of the physiological and enzymatic processes taking place in the grain and storage of the synthesized enzyme systems.
- Heat treatment and achievement of the necessary organoleptic qualities - color, aroma, taste.

The kilning process takes place in several interconnected phases. The first phase is called physiological and takes place when the temperature rises up to 40-45°C. The humidity decreases to about 30%, but the physiological processes (growth of the root; continuation of respiration, in which incomplete oxidation of carbohydrates takes place with the formation of ethyl alcohol and aldehydes) continue during the first phase. Besides, enzymatic processes continue and sugars and amino acids accumulate in the grain during this phase. The second phase is called enzymatic and it takes place in the temperature range of 45-70°C. In this phase, the humidity decreases to 10-12%, the vital processes in the grain slow down, but the hydrolytic enzymatic processes continue, as the enzyme systems are at their optimum. But since enzyme action depends on the water content, the enzymatic processes gradually slow down as the grain moisture decreases. The third phase is called chemical and it takes place at temperatures up to 105°C depending on the malt type produced. The grain moisture decreases to 3-3.5%, the enzymatic processes cease their action due to the inactivation of the enzymes at temperatures above 75°C. In this phase, a number of chemical processes continue, as a result of which melanoidins and other flavor malt components are formed (Narziss, 1999; Briggs et al., 2004; Kunze, 2004; Kreisz, 2009).

The kilning process is controlled by the temperature of the kilning agent and its flow, and no denaturation of the enzymes (when kilning at high temperature and humidity) or glazing of the malt, which occurs when using hot and dry air, should take place. The choice of initial temperature depends on the desired malt quality. Darker malts are characterized by higher degree of modification during germination in order to accumulate more sugars and amino acids that during kilning will form products of the Maillard reaction, i.e., more melanoidins. In general, kilning begins at about 50°C and with a gradual increase in the temperature up to 65°C, with the air being released into the atmosphere as it humidifies very quickly. The kilning process starts from the bottom layer and constantly develops through the layer. After 10-12 hours, the temperature above the layer rises and the relative humidity of the exhaust air decreases. This point is called "breakthrough" and indicates the earliest point (the moisture level in the top layer should not exceed 20%) when a slow rise in temperature to about 85°C is recommended. The color and aroma are obtained mainly by the Maillard reaction. The most important products of the Maillard reaction are melanoidins. They have major influence on the color and aroma, as well as on the pH and taste stability of the beer. A precursor of dimethyl sulfite (DMS) (S - methylmethionine) is converted to DMS. The enzymes are partially inactivated during kilning (Narziss, 1999; Briggs et al., 2004; Kunze, 2004; Kreisz, 2009).

The formation of additional malt qualities such as color, taste, and aroma is achieved through some additional operations such as roasting, smoking, and saccharification. Dark malts, acid malts, smoked malts, caramel malts are produced as a result of these operations. Each of these malts has a number of characteristics that give the beer or beverage characteristic qualities - color, taste, aroma (Narziss, 1999; Briggs et al., 2004; Kunze, 2004; Kreisz, 2009).

1.3. Main Processes in Beer Production

Beer production can be divided into 3 main stages – wort production, alcohol fermentation (main fermentation and maturation) and stabilization and finishing the beer (Kabzev and Ignatov, 2011).

Wort production is associated with the processes of milling, mashing, lautering and boiling of the wort. The purpose of the milling process is to release the components of the malt endosperm from the husk, which favors the subsequent process of mashing. Milling affects the sugar content of the malt mash, the yield of the extract, the lautering rate and the taste of the finished beer (Virkajarvi, 2001; Eaton, 2006; Kabzev and Ignatov, 2011).

The ground malt and the malt substitutes (unmalted barley, corn, rice, sorghum, etc.) are sent for mashing in mashing tuns. The aim of the mashing process is the conversion of high molecular mass substances into soluble forms which together with the available soluble substances, to pass into the wort. Enzymatic hydrolysis of biopolymers takes place during mashing: starch is hydrolyzed to mono-, di-, trisaccharides and dextrins; glucans - to oligosaccharides; protein substances - to peptides and amino acids. The degree of enzymatic hydrolysis and the wort quality are controlled by the so-called time-temperature pauses. In practice, different groups of mashing methods are used - infusion, decoction and combined methods. The aim of the method chosen is to achieve maximum yield of extract, as well as optimal physicochemical profile of the wort. Mashing lasts about 2 hours (Virkajarvi, 2001; Eaton, 2006; Kabzev and Ignatov, 2011).

The saccharified malt mash enters a lautering tun or a mash-filter, where lautering is performed. The wort is separated from the brewers' spent grains during the lautering process. The malt mash lautering takes place in two stages: separating the wort and rinsing the brewers' spent grains with hot water. The whole process takes 2-4 hours depending on the design of the lautering tun (Virkajarvi, 2001; Kabzev and Ignatov, 2011).

The wort and the rinsing water enter the brewing tun, where the wort is brewed with hops, as a result of which the composition is stabilized and the hop taste and aroma of the wort are formed. The following processes take place during boiling: dissolution and isomerization of hops bitter substances;

inactivation of enzyme systems; denaturation and coagulation of thermolabile protein substances; wort concentration; release of unwanted aromatic components (myrcene, dimethyl sulfide, etc.); melanoidin formation; pH lowering and others. The process takes 1-2 hours (Virkajarvi, 2001; Kunze, 2004; Briggs et al., 2004; Kabzev and Ignatov, 2011).

The hot wort is cooled to the fermentation temperature, and during the cooling process conditions for the separation of the sediment formed during boiling are created. The separation of the turb is conducted in two stages - hot turb separation and cold turb separation. The clarified and cooled wort is aerated in a stream and fed for alcohol fermentation (Virkajarvi, 2001; Kunze, 2004; Briggs et al., 2004; Kabzev and Ignatov, 2011).

Alcohol fermentation and maturation are the main stages in the formation of the beer taste and aroma profile. The main fermentation takes 3 to 6 days, and the maturation - up to 14 days (Branyik et al., 2005). In the fermentation process, yeast cells absorb carbohydrates, nitrogen and phosphorus substances and lipids from the wort, as a result of which alcohol and products of cell metabolism accumulate in the liquid and beer taste and aroma are formed (Virkajarvi, 2001; Kunze, 2004; Briggs et al., 2004).

Wort carbohydrates are the main carbon source for brewer's yeast. Two percent of the carbohydrates are used by yeast cells to form new biomass, and the remaining amount is transformed into alcohol and metabolic products. Brewer's yeasts absorb sugars in the following order: monosaccharides (glucose and fructose), disaccharides (sucrose and maltose) and trisaccharides (maltotriose) and they are fermented in the same sequence. The absorption of di- and trisaccharides begins after reducing the glucose concentration to 0.2-0.6% of the initial content. This process is known as catabolic repression. Small amounts of maltotriose are used to synthesize the reserve polysaccharides glycogen and trehalose (Shopska et al., 2016; Boulton and Quain, 2001; Kunze, 2004; Briggs et al., 2004; Russell, 2006; Tenge, 2009).

Nitrogen substances are used for the synthesis of both protein components and new cells. The wort amino acids are divided into 4 groups and the order of their absorption determines the formed secondary metabolites. Group A includes the amino acids that are assimilated the fastest - glutamine, glutamic acid, arginine, asparagine, aspartic acid, lysine, serine and threonine. The second group – group B - includes the amino acids that are assimilated immediately after the amino acids of group A - valine, methionine, leucine, isoleucine and histidine. Group C includes those that are most slowly assimilated - glycine, phenylalanine, tyrosine, tryptophan and alanine, and the last group – group D - includes the amino acids that are not absorbed, namely proline. Valine is necessary for yeast growth, but can be absorbed only after the amino acids of group A. Therefore, the cell begins to synthesize valine, forming α-acetolactate. The resulting metabolite is a precursor to diacetyl synthesis. The synthesis of precursors of aldehydes, higher alcohols and esters is also influenced by the order of amino acid absorption (Shopska et al., 2016; Tenge, 2009).

Wort lipids are used for cellular structures, or are involved in catabolic and anabolic pathways as regulators. Under anaerobic conditions, brewer's yeast is auxotrophic in terms of sterols and unsaturated fatty acids. Yeast can assimilate exogenous sterols, but only under anaerobic conditions when their synthesis in the cell is completed (Shopska et al., 2016; Briggs et al., 2004).

The yeasts transform the pyruvate into ethanol and CO_2 as a result of fermentation under anaerobic conditions. In addition to the two main metabolites, carbonyl compounds, higher alcohols, esters, organic acids and sulfur-containing compounds, which are related to beer taste and aroma, also accumulate. Therefore, it is necessary to know the biosynthesis and the regulation of the biosynthesis of yeast metabolites in order to regulate their amount within acceptable limits (Shopska et al., 2016; Boulton and Quain, 2001; Kunze, 2004). About 200 types of carbonyl compounds are found in beer, but the concentrations of aldehydes and vicinal diketones are of practical importance. Beer aldehydes originate from wort or are intermediates in the synthesis of higher alcohols from α-keto acids. The aldehyde with the highest concentration in beer is acetaldehyde, which is a precursor of ethanol. Aldehydes are synthesized mainly during the main fermentation and determine the "green" taste of the beer. The high temperature, the increased oxygen amount and the amount of the biomass inoculum, as well as the deteriorated physiological condition of the yeast cells are the reasons for the higher amount of acetaldehyde formed (Shopska et al., 2016; Briggs et al., 2004; Russell, 2006; Kabzev and Ignatov, 2011; Kunze, 2004). The vicinal diketones 2,3-butanedione (diacetyl) and 2,3-pentanedione, which have a low threshold of sensation in the final product, are of practical importance for beer quality. Vicinal diketones are an indirect result of yeast metabolism. Their precursors - α-acetohydroxy acids are intermediate metabolites of the biosynthetic pathway of isoleucine, leucine and valine. At the beginning of the main fermentation, part of the intracellular α-acetohydroxy acids are released in the fermentation medium, where they are converted into diacetyl and 2,3-pentanedione by oxidative decarboxylation (Boulton and Quain, 2001). Yeasts use alcohol dehydrogenase to reduce them during fermentation and maturation. The resulting end products - 2,3-butanediol and 2,3-pentanediol do not affect the beer organoleptic profile (Shopska et al., 2016). The formation of vicinal diketones is influenced by factors such as yeast strain, wort amino acid composition, aeration, pH, temperature, wort stirring (Shopska et al., 2016).

The second major group of metabolites is higher alcohols. More than 40 representatives of this group are found in beer - n-propanol, isobutanol, 2-methylbutanol (amyl alcohol) and 3-methylbutanol (isoamyl alcohol), as well as the aromatic alcohol 2-phenylethanol. Aliphatic alcohols enhance the alcohol taste and aroma and are the reason for the warming effect of beer, and the aromatic alcohol gives flowery aroma to beer and is a desirable ingredient. The major amount of these metabolites, over 90%, is formed during the main fermentation. These metabolites are synthesized by two metabolic pathways - catabolic (by deamination of the wort amino acids by yeasts) and anabolic (synthesized from the wort carbohydrates, i.e., obtained from pyruvate or acetyl-Co A as part of the aminoacid biosynthetic pathway). The choice of metabolic pathway depends on the wort amino acid content and the higher alcohol itself. n-propanol is produced anabolically because there is no corresponding amino acid. The catabolic pathway is chosen at higher amino acid content and with increasing chain length of the higher alcohols. Their synthesis and amount depends on the yeast strain, the wort composition, the fermentation conditions (Debourg and van Nedervelde, 1999; Shopska et al., 2016; Tenge, 2009; Boulton and Quain, 2001; Briggs et al., 2004; Russell, 2006).

Esters are characterized by a low threshold of sensation and a pleasant fruity-floral aroma. About 100 members of this group are found in beer. They give the beer different flavors - floral, fruity, herbal and others. These products are formed mainly during the main fermentation - up to about 60%, and the remaining amount is synthesized during maturation. Alcohols and carboxylic acids are required for ester synthesis. Although esterification is a chemical reaction, two enzymes are needed for the process to take place in the yeast cell: acyltransferase and estersynthetase, and the carboxylic acid must be activated with coenzyme A. The synthesis of esters in beer is influenced by the following factors – the yeast strain, the beer wort composition and, above all, the content of amino acids and sugars in it, the content of some metal ions in the wort, the fermentation conditions (Russell, 2006; Tenge, 2009; Shopska et al., 2016).

During fermentation, brewer's yeasts form about 110 organic acids, which lower beer pH. Some of them have characteristic taste and aroma that can undermine the beverage quality. Most of them are formed through the Krebs cycle. The amount of acids produced depends on the wort composition, the yeast strain used and the conditions of the fermentation process (Boulton and Quain, 2001; Shopska et al., 2016).

In addition to organic acids, fatty acids are also formed during fermentation, which usually degrades the quality of the beverage. Changes in the beer fatty acid composition occur during fermentation. The reason is the accumulation of higher fatty acids for lipid synthesis and the release of saturated short-chain fatty acids as by-products of this process, i.e., any change in fermentation conditions that stimulates yeast growth (increased temperature, aeration, and increased biomass amount) will lead to an increase in the concentration of these fatty acids (Boulton and Quain, 2001; Briggs et al., 2004; Shopska et al., 2016).

Beer quality is also affected by some sulfur-containing compounds. Some of them are present in the beer through the wort, and others are the result of metabolism - hydrogen sulfide, sulfur dioxide, dimethyl sulfide and mercaptans. Some of them are basic characteristics, for example hydrogen sulfide is characteristic of top-fermented beers. Their synthesis depends on a number of factors related to the wort composition, the fermentation conditions and the yeast strain (Boulton and Quain, 2001; Shopska et al., 2016; Kunze, 2004).

The last stage of fermentation is beer maturation. During this stage the last amount of fermentable extract is assimilated, the beer is saturated with CO_2, the unwanted metabolic components are reduced, the yeast cells are sedimented, the colloidal stability is improved, etc. (Shopska et al., 2016).

During the aging process, the beer cannot be sufficiently clarified. Its turbidity is determined by yeast cells and suspended particles. They are separated by lautering. The purpose of the process is to ensure the beer crystal clarity and its biological and colloidal stability. For this purpose, lautering additives, such as: kieselguhr, perlite, cellulose, cellulose charges and membranes, are used. The preservation of the beer organoleptic and physico-chemical properties for a certain period of time is called beer stability. Beer stability includes physicochemical (colloidal), biological and taste stability. Colloidal turbidity is due to the interaction of protein substances, polyphenols and small amounts of carbohydrates and trace elements. Precipitants (tannin and formaldehyde), adsorbents (bentonite), silica gel preparations, polyvinylpolypyrrolidone, enzyme preparations and antioxidants are used as stabilizing agents against them. Biological turbidity is most often caused by yeast and lactic acid bacteria, which cannot be completely eliminated during lautering. The beer is pasteurized in order to preserve the biological stability. The main sources of beer taste degradation are oxygen, carbonyl compounds, polyphenols, hop bitter substances and others. To prevent that, beer contact with oxygen should not be allowed and antioxidants should be used (Kunze, 2004; Kabzev and Ignatov, 2011).

2. Biological and Physiological Value of Malt and Beer

2.1. Phenolic Compounds and Antioxidant Capacity of Malt and Beer

Malt and cereals are essential for the beer organoleptic properties and its physicochemical characteristics. The fact that malt provides a complex of components with antioxidant capacity is important for the stability of the finished beer. Data in the literature show that the malt antioxidant capacity is largely responsible for the beer oxidative stability, which is primarily due to the large number of phenolic compounds in it. It has been found that about 80% of the total phenolic content of beer is due to malt and the remaining 20% to the hops used. In dark beer types, where dark and roasted malts are present, practically up to 95% of the antioxidant potential is formed by malt components. The formation of this potential mainly occurs in the processes of germination and kilning of malt. Kilning and roasting are crucial for the changes, and the products of the Maillard reaction during these two processes are important for the malt antioxidant potential (Carvalho et al., 2016; Vanderhaegen et al., 2006, Cortes et al., 2010; De Keukeleire 2000; Quifer-Rada et al., 2015; Cechovska et al., 2012; Leitao et al., 2012).

A number of compounds responsible for the antioxidant potential of the raw material have been identified in barley and barley malt - proanthocyanidin oligomers, hydroxycinnamic acid derivatives and low amounts of flavonols. They are found in free and bound forms, with the free and esterified fractions making a major contribution to the phenolic content. Catechin and ferulic acid are mainly found in malt. Malting is responsible for changes in the content of catechin, prodelfinidine B3, procyanidin B3 and ferulic acid, due to which differences in these components are observed in malt and barley. However, ferulic acid is resistant to the malting process and is therefore the main representative of the phenolic compounds in malt. In the malting process, a decrease in the free forms at the expense of an increase in the esterified representatives of the phenolic compounds, is observed. These changes are enzymatically induced. The phenolic acids in barley are derivatives of tyrosine and tyramine, their esters and glycosides, anthocyanides, proanthocyanidins, lignans and lignin components. They are found in the various structures of the grain - the husk, the pericarp and the aleurone layer. Ferulic acid predominates among the free phenolic compounds. These components undergo various transformations during malt production and are crucial for the beer colloidal stability (Carvalho et al., 2016; Salomonsson et al., 1980; Yu et al., 2001; Nordkvist et al., 1984; Hernanz et al., 2001; Holtekjolen et al., 2006; Madhujith et al., 2006; Lu et al., 2007; Dvorakova et al., 2008c; Dvorakova et al., 2008a; Magalhaes et al., 2011; Goupy et al., 1999; Samaras et al., 2005; Leitao et al., 2012; Inns et al., 2007, 2011).

Changes in the content of phenolic compounds in malt also occur in the stages of heat treatment of the grains - kilning and roasting. It has been found that high temperatures induce both degradation and polymerization processes of phenolic compounds. The data show that high processing temperatures are associated with a decrease in the content of ferulic acid. It is believed that in these cases the processes of melanoidin formation and the incorporation of phenolic compounds in the structure of melanoidins are responsible for this. In addition, thermal degradation of ferulic acid exterase has been observed, which is reflected in an overall reduction in the amount of phenolic acids in malt (Carvalho et al., 2016; Inns et al., 2011; Maillard and Berset 1995).

The heat treatment results in a non-enzymatic browning, known as the Maillard reaction. The reaction involves a cascade of successive reactions of interaction between the reducing sugars and the amino acids and the amino groups of the peptides and proteins, leading to the formation of a complex mixture of different compounds. The formation of different products depends on the temperature and the duration of the heat treatment, and the degree of change also depends on the residual moisture in the grain. It has been established that active pyrolytic processes and degradation of various components take place during roasting of malt and moisture of the grain up to 2%. In these cases, the color of the malt develops faster compared to "softer" heat treatment processes due to the formation of high molecular mass brown compounds. For this reason, there is a clear difference in terms of melanoidin content between light and caramel malts and roasted malts. Melanoidins have high reduction potential and intense brown color (Carvalho et al., 2016; Coghe et al., 2004, 2005, 2006; Yahya et al., 2014; Morales et al., 2005; Wang et al., 2011; Yahya et al., 2014; Cammerer et al., 2002; Carvalho et al., 2014).

Mainly polyphenols - catechins and ferulic acid, as well as products of the Maillard reaction are found in barley and malt. They play a major role in the malt antioxidant potential, and hence the resulting beer. These compounds play an essential role in the prevention of diseases such as cancer, cardiovascular diseases and others. The potential of phenolic acids is due to their ability to donate hydrogen and electrons, as well as to form stable complexes with radicals. In general, flavonoid phenolic compounds have been found to have greater potential. The data in the specialized literature show that the formation of the malt antioxidant potential occurs mainly in the processes of kilning and roasting. This is due to the fact that during the two heat treatments polyphenols are released and reductones and products of the Maillard reaction accumulate (Carvalho et al., 2016; Samaras et al., 2005; Vanderhaegen et al., 2006; Inns et al., 2011; Landete, 2013; Rivero et al., 2005; Maillard and Berset 1995; Subba Rao and Muralikrishna 2002; Zhao et al., 2010; Coghe et al., 2003, 2006).

The formation of the antioxidant potential of the grain raw materials, as well as of the malts obtained from them, begins with the stage of steeping. It was found that the content of phytic acid, which has certain negative properties on the wort biological properties, is reduced in the process of steeping. Reducing the concentration of this component significantly improves the nutritional properties of cereals and products derived from them. The process of germination is also accompanied by an increase in the nutritional value of cereals due to the synthesis of many functional bioactive components such as antioxidants, vitamins and others. Germinated grains have a higher content of certain vitamins than non-germinated grains. Germination also affects the dietary fiber content. The amount of β-glucan decreases during germination, but the concentration of water-soluble arabinoxylans increases (Carvalho et al., 2016; Greiner and Konietzny,

2006; Albarracin et al., 2013; Hassani et al., 2016; Rao and Muralikrishna, 2006; Krahl, 2010).

The processes of beer production are also important for the biological value of the beer produced. For example, mashing increases the concentration of some of the components, as enzymatic hydrolysis of the raw material and bringing some components to a water-soluble form is allowed to happen. The degree of extraction of these components, especially the polyphenols, depends on the ratio of malt to water when mixing the malt mash. The changes in the antioxidant capacity are also determined by the processes of brewing, fermentation and storage of the finished beer (Shahidi and Naczk, 2004). Data on the content of polyphenols in different beer types can be found in the specialized literature. For example, the polyphenolic content of lager beer types is about 30-40 mg/100 ml, while in dark beer types it is 1.5 times higher. The low-alcohol beers available on the market are characterized by about 2 times lower content of phenolic compounds than the lager beer types. Seventy-eight different phenolic compounds have been found in beer, 81% of which are proanthocyanidins, 14% are phenolic acids and 5% are flavanoids. Proanthocyanidins have low biological activity in the gastrointestinal tract, but can be fermented by the microflora there and provide the necessary antioxidants. The phenolic acids – ferulic acid, gallic acid and sirinic acid - are found in free and bound form. The main representatives of flavonoids are catechin, epicatechin and quartcetin. All these components have an antioxidant potential that is many times higher than that of vitamin C and vitamin E. Most authors report three times higher values of the indicators, depending on the system used to determine the antioxidant activity. Beer polyphenols have been found to be bioavailable to the human body, increasing the antioxidant capacity of blood plasma within 30 minutes after consumption. It is important to note that the bound forms are difficult to digest, but instead they are fermented by the microorganisms in the column and thus become bioavailable to humans (Fulgencio et al., 2009; Shahidi and Naczk, 2004; Gerhäuser, 2005; Nardini and Ghiselli, 2004; Sánchez-Moreno et al., 1998; Villaño et al., 2005; Ghiselli et al., 2000).

The melanoidin biological value has also been well studied by various authors. In general, dark beers contain about 2 to 2.5 times more melanoidins than lager beer types. From nutritional point of view, melanoidins provide different mechanisms of antioxidant action and do not show cytotoxic action. However, data in the literature show that melanoidins have a much lower potential than some phenolic components (catechin and quartcetin) (Fulgencio et al., 2009).

It is well known that the consumption of plant foods that are rich in antioxidants reduces the risk of certain diseases. This in turn makes the application of these products interesting in the development of diets providing complex matrices rich in bioactive substances and their combination with certain types of microorganisms, provides a balanced diet and reduces the risks associated with the development of some modern diseases. Beer consumption provides a balanced set of components with antioxidant capacity, and between 5% and 10% of antioxidants may be due to beer and its components depending on the overall food profile (Fulgencio et al., 2009; Serafini, 2006; Woodside et al., 2005).

2.2. Biological Activity of Hop Substances

Hop components also have antioxidant potential. The main potential of hop preparations is due to the content of polyphenols, which can prevent oxidative stress, which is associated with cancer and cardiovascular disease. It has been found that hops and hop extracts rich in phenolic compounds inhibit the oxidative breakdown of lipids, have radical scavenging and metal-reducing ability. Humolons also have certain antioxidant and biological activity. They have been found to lead to positive effects in patients with osteoporosis, leukemia and other diseases. Despite most studies, their mechanism of action is not yet fully established, and their percentage in relation with malt components is relatively low (Tobe, 2009).

3. TAILOR-MADE CONCEPT FOR THE PRODUCTION OF BEER WITH INCREASED BIOLOGICAL VALUE

By definition, a tailor-made product is a product that is designed, adapted or situated to achieve a purpose, and in this case for consumption by a group of people.

The data presented so far provide convincing evidence that beer and its raw materials have biological significance for the human body. Beer is a beverage that can be subjected to modeling and optimization of the composition in order to fulfill tailor-made purposes. However, this requires a conceptually new approach that is related to the 5 directions of developing a functional product (Siro et al., 2008):

- Elimination of ingredients from food that are known to have an adverse effect on the body;
- Increasing the concentration of the constituents in the food to the amount which will show the beneficial effects expected or increasing the concentration of the non-nutritional ingredients to a level known to have beneficial effects;
- Addition of components that are not present in most foods and which are not nutrients but have a proven positive effect on the body;
- Replacement of components whose uptake is excessive and leads to adverse effects, with those with beneficial effects identified;
- Enhancing the bioavailability or stability of a component known to have a positive effect or reduce the risk of developing a disease.

The choice of the grain raw material (malt or unmalted product) is related to the specifics of the produced beverage and the desired organoleptic profile. One has to take into account some specifics of the industry, which uses mainly barley and barley malt, wheat and wheat malt, as well as substitutes such as unmalted barley, corn, rice and others. This is necessary due to the specifics of production and, last but not least, the search for opportunities for production of products with normalized cost. The choice of these raw materials is determined by their high starch content, which, however, is not in accessible form for the technological process. Therefore, it is necessary to go through a malting process, which aims to improve the physicochemical composition of the grain and only then to use the grains in the technological process for beverage production. The malting process, as already seen, also improves the biological value of the product obtained and is therefore an invariable step in the concept of developing new types of beer with functional characteristics. Mono- and disaccharides in cereals absorbed by microorganisms (glucose, fructose, and sucrose) are not more than 3% of the dry matter, which requires measures to increase the content of these components in the extract. It is necessary to know whether the cereal has been malted or not, as there are differences in its carbohydrate composition. Enzymatic hydrolysis during cereal malting leads to an increase in the amount of degradation products. Carbohydrates in wort and beer are mainly products of starch hydrolysis - dextrins. The rapid degradation of straight-chain dextrins by pancreatic α-amylase releases energy, while the degradation of branched-chain dextrins by oligo-1,6-glucosidase is slower and they can be used by the intestinal microflora as a nutrient source (Kabzev and Ignatov, 2011). The application of malted raw materials for wort production is an indispensable step in a tailor-made concept. However, since the production of small portions of malt with special characteristics is economically unprofitable, our proposed concept is based on optimizing the wort composition by using malts, which are available commercially. Our concept includes the following basic steps shown in Figure 1. The first step involves the characterization of the main raw material - malt. For this purpose, a wide range of malts should be selected to be characterized in terms of their physicochemical (extract, yield of extract, pH, diastatic power, color, etc.) and biological (phenolic compounds content, antioxidant potential) characteristics and if possible, to look for a correlation of the received data between the two groups. Single-factor experiments and standardized methods can be used to determine these characteristics. Then, a database is accumulated and it can be used to perform the selection for the next step.

This immediately raises the question of how to make the malt selection in view of the wide profile of characteristics of the individual types. As already mentioned, mainly barley and wheat malt are used in the brewing industry, with the starch content being the leading parameter. Our results and the modes used for malt production show that malts can be divided into three main groups - basic, special and functional. This classification is also used by malt producers. The basic malt types are those that are used in the highest percentage in the composition of the ground raw material mixture to form the wort, while the special and the functional malts serve to give taste, color, aroma and some other wort functionalities. The basic malt types provide the necessary wort extract, and the yield of the extract should be in the range of 80-85%, which guarantees the quality of the resulting beer. These malts have high proportion of active enzyme systems that are used in the mashing process, can be used in combination with the other two malt groups and give the main profile of the final beverage. Usually their amount in the total malt ground raw material mixture varies in the range of 40-80%. This group includes malts such as Pilsen, Viennese, some types of Munich caramel malts and others. They have high starch content (over 60%).

Special malts provide the beverage with special characteristics such as color, taste, and aroma. They are kilned under special regimes or have undergone additional heat treatment in order to form their special properties. As these malts undergo intensive malting, kilning and, in some cases, additional heat treatment, they form lower extract and, thus, lower extract yield. In general, the increase in the degree of heat treatment in this group of malts leads to a gradual decrease in their brewing characteristics - starch content and extract yield and increase in the wort color intensity. The reduction of the brewing characteristics determines the reduction of their percentage participation in the malt ground raw material mixture, as they usually participate in an amount of up to 30-40% (and for some malts this percentage is about 5-10%). Functional malt types have the task to improve one or more wort qualities - pH, aroma, in some cases color, diastatic power. They are usually placed in the wort mixture in a certain amount, which is determined on the basis of the desired functional effect. These malts have quite similar parameters to those of the basic malt types, but are used in an amount of 5-10%.

In the tailor-made concept presented by us, the choice is based not only on the brewing malt characteristics, but also on its biological value. The division of malts into three groups - basic, special and functional - is an appropriate method for malt selection in terms of their biological activity. The biological activity can be determined by their phenolic compounds content and antioxidant potential.

The basic malt types are characterized by low to medium content of phenolic compounds and lower color intensity. This in turn determines lower antioxidant potential. This reduced potential is largely offset by the special malts group. They are characterized by a wide range of phenolic reserves, due to the changes we have already talked about in the section on malt production. With regard to the antioxidant potential of special malts, there is a clear trend - an increase in antioxidant capacity with increasing the malt color intensity. The basic malt types show low activity due to the low content of phenolic compounds. As the degree of heat treatment and the malt color increases, so does its antioxidant capacity. This immediately leads to the conclusion that the biological potential of the malt and the corresponding wort is due not only to the phenolic compounds, but also to the products of the Maillard reaction. In this area, too, the functional malt types have closer characteristics to those of the basic malt types.

The data given so far show that there are two opposite tendencies in the combination of malts in the ground raw material mixture - the high wort extract is provided by basic malts, while the wort biological value, and hence the beer biological value is provided by the higher potential. It is obvious that by classical one-factor experiments it is difficult to find the optimal component composition of the malt ground raw material mixture in order to obtain wort with pronounced biological value. This is the second step of the tailor-made concept - modeling the wort composition through a suitable mathematical and statistical apparatus (Figure 1).

Tailor-Made Concept for New Beer Types ... 33

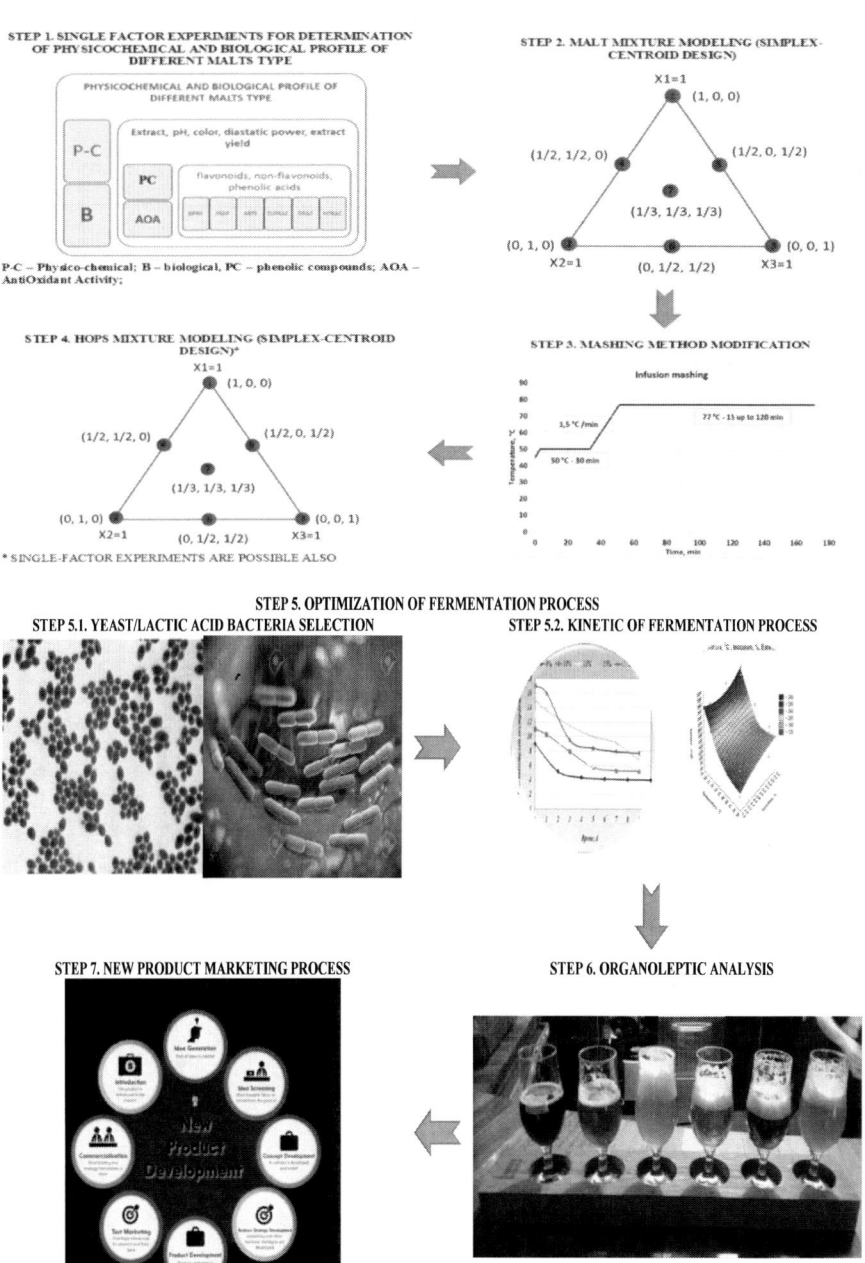

Figure 1. Tailor-made concept steps.

The choice of a statistical technique for modeling the composition of the malt mixture is an important step in the tailor-made concept. For this purpose, statistical techniques are used to design the experiment. This is a systematic approach of selection of factors that have a significant impact on a particular objective function, as well as their levels of variation. The design of the experiment is a powerful tool for adapting experimental data to empirical dependencies and for providing information about a given system action. The design of the experiment can be used for various purposes, such as optimizing the accuracy of the estimation of the model parameters (e.g., D-optimal plans) or optimizing the accuracy of model prediction (e.g., V-optimal plans) (Buruk Sahin et al., 2016).

The experiment planning process involves a new combination of steps. The first step is to determine which factors and at which levels have significant influence on the target function. The second task is to decide whether the resulting model will be used as an optimization function. The group of design of experiments itself is quite large and the researcher must decide which method for experiment planning to use. The experimenter must also identify both the independent factors and the reaction factors of interest to him or her. Permissible levels of independent factors must be determined and this determines the experimental area. If a polynomial model is used, its degree should be specified, and if a more fundamental model is to be used, it should be more specific (Buruk Sahin et al., 2016).

In our case it is necessary to model the composition of a mixture of several malt types, which to some extent defines the type of the planned experiment used - modeling of mixtures. This is a special approach in which the individual components of the mixture are varied, but the total amount is always 1 (100%). This method is applied in various industries - pharmaceutical, polymer, ceramic and makes it possible to select the optimal composition of the mixture in order to achieve an optimal profile of the resulting product or article (Buruk Sahin et al., 2016).

Mixture modeling is a special type of design of experiment in which the varying factors are mixture components, and the target function is affected by the different proportions in the mixture. The sum of the proportions of the mixture must be equal to 1. In this case, the standard modeling methods cannot be applied. The equation of a mixture of q components is shown in equation (1) (Buruk Sahin et al., 2016):

$$0 \leq x_i \leq 1; \quad i=1, 2, ..., q; \quad \sum_{i=1}^{q} x_i = 1 \qquad (1)$$

where: x_i - amount (proportion) of the i-th component in the mixture; q-components form a simple (q-1)-dimensional simplex.

The modeling itself can be done through three different methods. The first is the so-called Simplex-Lattice design. This is one of the most commonly used methods for modeling of mixtures and is defined as follows: The "$\{q, m\}$" grid for q components consists of points defined by the following coordinate settings: the proportions of each component are assumed m-1 equally spaced values from 0 to 1 (Buruk Sahin et al., 2016):

$$x_i = 1, \frac{1}{m}, \frac{2}{m}, ..., \frac{m-1}{m}, 1; \quad i=1, 2, ..., q; \qquad (2)$$

using all combinations in proportion. The number of experimental points in the "$\{q, m\}$" grid is as follows:

$$\frac{(q+m-1)!}{m!(q-1)!} \qquad (3)$$

With the Simplex-Centroid design plan, all components vary in the same way (between 0 and 1) and no factor space restrictions are introduced. The central point of the plan, which includes all components, is always implemented. This method is suitable for mixtures with up to 4 components (Buruk Sahin et al., 2016).

There are other types of plans that are derived from the two types we have described so far. They include restrictions on the proportions of the component x_j, which varies in the range of L_j to U_j. The general form of constraints is given by the dependence:

$$\sum_j x_j = 1;\ 0 \prec L_j \leq x_j \prec\sim U_j \tag{4}$$

I-, D- and G-optimality are widely used for problems with limited mixtures. I-optimality focuses on minimizing the average scaling up of the prediction for the difference in the experimental area. G-optimality focuses on the variance of the general prediction equation. The D-optimal design focuses on estimating the best possible coefficients of the model. This can be achieved by minimizing the determinant of $(X'X)\ -1$, where X is the matrix of the appropriate proportions of the component and possibly cross-products between the proportions, depending on the model, and X' is the transposition of X (Buruk Sahin et al., 2016).

As can be seen, the modeling of a mixture of several malt types can be used to realize the tailor-made concept. In this case, a combination of malts should be selected, one of which must be from the basic malts group. This malt will determine the type of the beer obtained - lager, wheat, Viennese, Dortmund and others. Then special malts are selected. The specific choice of components to enter the mixture can be made on the basis of empirical and experimental knowledge of the organoleptic characteristics that each malt type would give the beer, the malt phenolic capacity, the malt antioxidant potential. The number of components in the mixture, as well as the possibility of limiting the amount of one or more of them determines the type of modeling method used. Restrictive conditions can be both numerical and organoleptic evaluation of the final beverage. The implementation of the experiments at the laboratory level, the development of models, their interpretation and optimization are conducted at this point of the concept (Figure 1). As a result of this step, the optimal composition of the malt ground raw material mixture is determined, and it can be used for the production of beer with increased biological value.

The malt phenolic compounds undergo significant changes in the malt obtaining process and especially in the mashing process. The method of malt grinding is of paramount importance. Data by *Szwajgier, 2011* show that wet milling leads to a reduction in the ferulic acid content and the total phenolic compounds content in the wort. This applies to malts whose husks are rich in phenolic compounds. The content of phenolic compounds is significantly influenced by the mashing method chosen - infusion or decoction. Free cinnamic acid is dissolved during mashing by extraction and enzymatic release (Vanbeneden et al., 2007). The temperature optimum for cinnamol esterase action has been found to be about 30°C (Vanbeneden et al., 2008b), but its release is also associated with the synergistic action of some arabinoxylan hydrolyzing enzymes (Bartolome et al., 1996). The temperature optimum for the release of ferulic acid is 40°C, but at this temperature the activity of cinnamol esterase decreases by 40% within 60 minutes. These data are confirmed by other authors who found that at 40-45°C the best extraction of ferulic acid is observed (Coghe et al., 2004). In this case, we can suggest that the data for the basic malt types confirm the data cited by other authors. The extraction of phenolic compounds in mash depends on the mashing time, the mash concentration and the amount of the malt ground raw materials mixture. More phenolic compounds (ferulic acid) are released at finer milling and at more concentrated mashes (Vanbeneden et al., 2008b). Other data show that the longer mashing, i.e., incubation of the mash at lower temperatures leads to higher ferulic acid content, and this tendency is valid up to temperatures of about 60°C (Schwarz et al., 2012b). After this temperature, ferulol esterase is completely inactivated. The pH optimum of the enzyme's action is 5.8 (Vanbeneden et al., 2008b). The authors found that only a small portion of hydroxycannellic acid is extracted in the mash, and the bulk of it remains in the brewers' spent grains (Vanbeneden et al., 2007). In the classical infusion mashing method, it was found that the total amount of phenolic compounds increases up to a temperature of 78°C, followed by a decrease in the phenolic content of the mash (Zhao & Zhao, 2012). The mashing method is also essential for the yield of phenolic compounds. The data show that decoction methods provide higher yield of polyphenols in the wort (Jurkova et al., 2012).

The mashing method is also crucial for the beer alcohol content. Its modeling in a way that allows both the ethanol content of the final product to be reduced and the biological value of the beer to be ensured is an important step in the development of the tailor-made concept. Skipping the pause at 63°C is an important step in reducing the content of fermentable sugars (Ivanov et al. 2016). As can be seen, this pause is not essential for the phenolic profile of the wort. As a result of such a modified mashing method, wort with a profile of fermentable sugars that allows the reduction of the alcohol content and the production of a low-alcohol beverage, can be obtained. Of course, it is possible to choose the production of classic beer - with normal alcohol content for this beverage. The approach of reducing the content of fermentable sugars allows the wort to be used for conceptually new beverages in which brewer's yeast can be replaced by probiotics such as *Saccharomyces boulardii* or some types of probiotic lactic acid bacteria and to market new types of beverages based on wort. Strictly speaking, they cannot be classified as beer, but can acquire functional value, both on the basis of their physicochemical composition and on the basis of their probiotic characteristics. The use of aqueous extracts of malted cereals (such as wort) as a substrate for the growth of various types of lactic acid bacteria leads to the production of a functional product. Strains of *L. fermentum, L.reuteri, L.acidophilus, L. plantarum*, which can assimilate maltose, sucrose, glucose, fructose and free amino nitrogen accumulated in the wort, are used for fermentation (Charalampopoulos et al., 2002). The addition of cereal extract improves the survival of probiotic cultures in the gastrointestinal tract. Fermentation of cereals such as oats, corn and wheat under the action of probiotic microorganisms is useful because it provides better digestibility of food. This is associated with a reduction in the amount of indigestible carbohydrates such as polysaccharides and oligosaccharides, the presence of the vitamin B complex and the release of minerals such as calcium, zinc, ferrum and others (Soccol et al., 2010). There is evidence in the literature that the survival of lactic acid bacteria is significantly improved by the use of barley malt as a basis for beverage production (Charalampopoulos et al., 2003).

Wort contains fiber, which is obtained as a result of hydrolysis of the cell walls of barley in the process of malting and mashing and from there it passes into the wort and beer. High molecular weight β-glucans can cause malt lautering problems, but low molecular weight soluble glucans are good for digestion because they contain a β-bond. In addition to contributing to the healthy functions of the colon, they can lower blood cholesterol, thus reducing the risk of coronary heart disease. Wort and beer are a source of a number of vitamins, mainly those of group B, and they come mainly from malt. High content of the bioavailable form of silicon has also been found in wort and beer. This element is essential for the synthesis of healthy bones and plays an important role in reducing the levels of aluminum in the body (Kabzev and Ignatov, 2011).

The increase in the biological value of wort is also due to the inclusion of hops in it. As we have already seen, hops components - acids, essential oils and humolons - have certain biological significance for beer, and hence for the human body. Different types of hops are available on the market and their combination is rather an empirical process in which a certain taste and aroma of the beverage are sought. An approach with modeling the composition of the mixture is also possible here, but since there is certain instability of the hop components, it would rather complicate the process of developing the new beverage. In general, we can conclude that the inclusion of hops can increase the biological value of the beverage. Here is the place to note the existence of an interesting contradiction. The phenolic compounds of malt and hops are of particular importance for the beer colloidal stability. The data show that the phenolic compounds of hops are largely responsible for the problems with colloidal stability, as they are more reactive to the beer protein components. The use of essential oils in the brewing industry is not a new direction. It is a well-known fact that hop essential oils are increasingly used in the production of beer, which on the one hand retain the typical hop taste and aroma, and on the other hand, does not lead to colloidal destabilization of beer. This is due to the fact that only oil components without phenolic compounds responsible for the colloidal changes in the beverage, are extracted selectively in the production of essential oils. Another well-known fact is that some beer consumers do not like the hop taste and aroma of the beverage. It is therefore necessary to look for opportunities to replace hops with other types of flavors and tastes to attract new consumers of beer and beverages based on beer and wort. At this step in the concept (Figure 1) it is necessary to select essential oils, which not only give the beer a new taste and aroma profile, but also allow an increase in its biological value. It is known that oils have antimicrobial capacity so at this stage of the concept it is necessary to determine their influence on the microorganisms used for wort fermentation.

The production of fermented foods, beverages and other fermentation products is based on a number of processes. Alcohol fermentation is a major process in which the beer quality is formed. It is a biochemical process that is carried out by yeast cells, in the vital activity of which the substrate – wort - is transformed into a product with a specific organoleptic profile, qualitative and quantitative composition (Boulton, 2006; Mendes et al., 2017; Lengyel and Panaitescu, 2017). As a result, the substrate is absorbed, biomass and metabolic products accumulate, and the taste and aroma profile of the beverages is formed. In most cases, fermentation is the time determining process in production and knowledge of its dynamics leads to the urgent search for opportunities to optimize production, reduce production costs and increase the productivity of fermentation systems. Here are different approaches for modeling the process in order to form the organoleptic profile we are looking for. Perhaps the most modern approach is to combine the determination of the fermentation process kinetics with the analysis of the metabolic flows (Antalick et al., 2015; Boulton, 2006; Mendes et al., 2017; Lengyel and Panaitescu, 2017). Research is in the field of modeling and identification of the kinetics of different types of fermentation processes, used in the food and biotechnology industries. Thus, a comparative analysis of the obtained original research results with the research of other authors can be made.

The processes taking place in the microbial population and in the technological equipment intended for this purpose, respectively, depend on a number of factors - pH, temperature, hydrodynamic conditions in the devices, etc. (Kostov, 2015). Each fermentation process can be described by a system of three basic differential equations that take into account the changes in the biomass concentration, the metabolic products concentration and the changes in the substrate over time:

$$\left| \begin{aligned} \frac{dX}{d\tau} &= \mu X \\ \frac{dP}{d\tau} &= qX \\ \frac{dS}{d\tau} &= -\frac{1}{Y_{X/S}}\frac{dX}{\tau} - \frac{1}{Y_{P/S}}\frac{dP}{d\tau} \end{aligned} \right. \qquad (5)$$

where: X - biomass concentration, g/dm^3; P - product concentration, g/dm^3; S - substrate concentration, g/dm^3; $Y_{P/S}$, $Y_{X/S}$ - yield coefficients; μ - specific growth rate, h^{-1}; q - specific rate of accumulation of the product, g/(g.h);

This system can be supplemented by other equations for each specific case, for example for the changes in the secondary metabolites - esters, aldehydes, higher alcohols, vicinal diketones, as is the case with beer production. The specific type of the system of differential equations is determined by the researcher's needs and the type of fermentation process studied (Bailey and Ollis, 1986). Secondly, the system of differential equations can be built both for the whole process and for its parts - lag phase, exponential phase, stationary phase, and phase of death of the microorganisms. The main task in modeling the kinetics of any fermentation process is the identification of parameters in the system of differential equations. Different methods are used for this purpose. Approaches to the analysis of microbiological systems are classified according to the number of components used to describe the cells, as well as whether the population is heterogeneous or consists of averaged cells (Bailey and Ollis, 1986; Biryukov, 2004). The modeling of the fermentation process itself is a significant step that requires a different approach in the presented concept. But the development of a complete concept for beer modeling includes knowledge of fermentation kinetics, but also the kinetics of loss of biological potential due to the formation of complexes in beer.

The assessment of the antioxidant capacity of different beer types shows that their antioxidant capacity largely depends on the beer color (Polak et al., 2013). In fact, the higher antioxidant activity of malts is mainly due to the products formed in the Maillard reaction, which correlates with the malt color and the melanoidin content (Chandra et al., 2001; Coghe et al., 2003). It has been shown in the literature that malt extracts obtained with 80% acetone exhibit strong *in vivo* and *in vitro* AOAs, and can bind hydroxide and superoxide radicals (Qingming et al., 2010). Catechins, caffeic acid, ferulic acid have a major contribution to the antioxidant potential (Leitao et al., 2011). Catechin, for example, has a higher metal-reducing ability, while ferulic acid has a higher DPPH radical inhibiting ability (Lu et al., 2007).

4. SUSTAINABILITY OF BEER PRODUCTION – BREWERS' SPENT GRAINS AS AN ELEMENT OF FUNCTIONAL FOODS

An important feature of modern production is its sustainability and the possibility of application of waste products in other areas of life, or even their application to obtain new types of food. In this respect, the brewing industry is no different from other industries. The main waste products from the brewing industry are brewers' spent grains (BSG). They form 85% of all waste products, but their characteristics meet the conditions for sustainable production and allow application as a functional element in the food industry. They are lignocellulosic materials whose chemical composition (protein, fiber, phenolic compounds, etc.) makes them extremely suitable for the production of functional foods (Aliyu and Bala, 2011; Mussatto et al., 2006; Ivanova, 2018).

4.1. Chemical Composition and Characteristics of Brewers' Spent Grains

The BSG physico-chemical profile varies depending on the barley variety used for malt production, the malting conditions and the wort production conditions. The data show the following general profile of BSG (% of dry matter): cellulose - 16-25%; hemicellulose - 28-35%; lignin - 7-27.8%; ash - 2.3-8%. Separate studies show that BSG contain 79.9%

carbohydrates, between 2.5% and 11% lipids, 3.3% crude fiber (Aliyu and Bala, 2011; Mussatto et al., 2006; Ivanova, 2018; Kanauchi et al., 2001; Russ et al., 2005; Mussatto and Roberto, 2006; Mussato et al., 2008; Adeniran et al., 2008; Khidzir et al., 2010).

Data from various microscopic preparations show that BSG contain a number of fibrous tissues composed mainly of arabinoxylan, lignin and cellulose (Aliyu and Bala, 2011).

BSG also contain minerals (calcium, cobalt, cuprum, ferrum, magnesium, manganese, phosphorus, potassium, selenium, sodium and sulfur and all in concentrations below 0.5%), vitamins (biotin, choline, folic acid, riboflavin, thiamine and others) and amino acids (valine, alanine, serine, glycine, glutamic acid and aspartic acid, tyrosine, proline, threonine, arginine and lysine) (Aliyu and Bala, 2011; Mussatto et al., 2006).

In their works Ivanova, 2018 and Ivanova et al., 2016 conducted a study of some of the more important physicochemical characteristics of BSG obtained from the so-called basic malt types (Vienna, Pilsen, Munich, Wheat), as well as a special malt type - Caraffa. The data show that the BSG contain between 5% and 30% starch, the content depending on the malt type. The starch contained is highly gelatinized, with the degree of gelatinization varying from 35% (for light malts) up to 60% (for malts that have undergone higher heat treatment). The latter is essential as it allows BSG to be classified as a raw material that is suitable for extrusion.

As already commented, polyphenols are one of the sources of biological activity of beer, as they have significant antioxidant potential. In their studies Ivanova, 2017 and Ivanova et al., 2016 showed that the polyphenol content in the studied malt types varies between 130-200 mgGAE/100 g, and the content depends on the malt type used. These results are in agreement with the results of other authors (Moreira et al., 2013; McCarthy et al., 2012; Dvořáková et al., 2008; Magalhães et al., 2011), which show that the BSG phenolic profile is influenced by the raw material from which the malt is obtained, as well as the production regimes. Another important feature is that the products of the Maillard reaction have the ability to retain polyphenols in their structure, which is why some darker malts and their

BSG are characterized by lower phenolic content (Inns et al., 2011; Maillard and Berset, 1995).

The BSG antioxidant potential is important for their quality. Ivanova, 2017 and Ivanova et al., 2016 show that BSG have significant antioxidant potential. It varies between 400-600 µmol TE/100 g, depending on the method of its determination and the evaluation system used - DPPH or FRAP. In all variants studied, the metal-reducing ability is higher than that of the DPPH radical inhibition, which should be explained by the antioxidant profile of the BSG components. The high antioxidant activity of Caraffa malt is due to the significant amount of melanoidins that are formed during its roasting. Melanoidins are known to have antioxidant capacity.

Additional studies of BSG obtained from different malt types show that the malt type (light, dark, roasted, smoked) is essential for the content of polyphenols and antioxidants. The data from Ivanova et al., 2020 show that the phenolic content reaches 2000 mg/dm^3, with an increase in the phenolic content being observed with an increase in the degree of heat treatment. The antioxidant content is similar, as the increase in the BSG capacity is associated with an increase in the degree of processing and to a large extent in the content of products from the Maillard reaction. Similarly to previous studies, the metal-reducing ability is higher than that of inhibiting the DPPH radical.

BSG also contains a significant amount of crude fiber. This is the residue which is obtained after treatment with solutions of acids and alkali with a certain concentration under known conditions. Fiber is important for human nutrition, providing conditions for the proper growth of microorganisms in the gastrointestinal tract, acting as a prebiotic and more. The fiber content is up to 35% and is usually up to 2 times higher than the fiber content in the malt from which the BSG are obtained. Another important feature is the fact that dark malt types contain lower amounts of fiber (Ivanova, 2018; Ivanova et al., 2016).

4.2. Extrusion of Brewers' Spent Grains as a Method for Their Preservation and Increase of Their Functional Properties. Application of Brewers' Spent Grains in Tailor-Made Foods

4.2.1. Extrusion of Brewers' Spent Grains as a Method for Their Preservation and Increase of Their Functional Properties

Immediately after lautering, the BSG contain up to 81% water. This, together with the fermentable sugars they contain make BSG a rather unstable product (Aliyu and Bala, 2011; Mussatto et al., 2006; Huige, 1994; Russ et al., 2005).

In general, the methods used to preserve BSG are energy-intensive and make them relatively unusable in industry. Drying in industrial conditions is the most effective method for preserving BSG. However, this method is very energy-intensive and therefore many breweries do not practice it. Kilning as a preservation method has the advantages of reducing the product volume as well as the transportation and storage costs. Many breweries use two-step kilning: first, they reduce the moisture content below 60% by pressing and then kilning to provide moisture content below 10% (Santos et al., 2003). Traditional methods for kilning of BSG are based on the use of drum dryers. *Bartolome et al., 2002* studied the possibilities of preserving BSG by freeze-drying or freezing. These preservation methods reduce the product volume without affecting the composition, while freezing is an inappropriate method as it affects the content of certain sugars, such as arabinose (Santos et al., 2003; Bartolome et al., 2002; Mussatto et al., 2006).

A promising method for preserving BSG is the application of extrusion. This, in general, is the process of leaking a product through an opening under the action of applied pressure. In this case, the product is subjected to a combined effect of heat, moisture, pressure and shear stresses. Thus, the material undergoes various changes - homogenization, plasticization, gelatinization, structural changes and others (Ivanova, 2018; Penov and Petrova, 2014; Guy, 2001; Riaz, 2000). Over the years, the process has become necessary in the processing of cereals such as corn and wheat to obtain a number of food products. The method has a number of advantages (Ivanova, 2018; Penov and Petrova, 2014; Guy, 2001; Riaz, 2000): it has high productivity; it provides an opportunity for quick changes in the used recipes; the product is sterilized by undergoing thermal exposure, yet retaining to a significant extent its nutrients; the products obtained from the processing have a high degree of absorption due to starch gelatinization, protein denaturation, degradation of trypsin and pepsin inhibitors and destruction of the enzyme systems of the raw materials; some of the products boil easily, which leads to energy savings; the method allows the production of staple foods; the number of technological operations is reduced, etc. Last but not least, the method is characterized by energy savings.

Three types of extrusion are used in industry – cold extrusion, hot extrusion at low pressure and hot extrusion at high pressure. The first method is used to make pasta, while the second and third methods are used in the production of a number of foods. Regarding the hardware layout, two types of systems are used - single-screw and double-screw extruders (Ivanova, 2018; Penov and Petrova, 2014; Guy, 2001; Riaz, 2000).

The single-screw extruder is characterized by three operation areas - power supply; area of mixing; area of shearing. As a result, shear stresses on the material are observed, which leads to the conversion of mechanical energy into heat. The mixing takes place in the first zone. The second zone is characterized by the plasticization of the product. The third zone is characterized by shear stresses and intensive homogenization and plasticization of the product. This is due to the different pitch and diameter ratios of the extruder in the different zones of the system. This extruder type requires pre-homogenization of the processed mass, as well as optimization of the moisture and the fat content in the mixture (Ivanova, 2018; Penov and Petrova, 2014; Guy, 2001; Riaz, 2000).

The second type of systems are the so-called double-screw extruders. They have a more complex construction, but greater flexibility and better management at the same time (Ivanova, 2018; Penov and Petrova, 2014; Guy, 2001; Riaz, 2000; Akdogan, 1996).

The extrusion process is characterized by changes in the physicochemical composition of the products, which in most cases increase their biological significance (Ivanova, 2018; Penov and Petrova, 2014; Guy, 2001; Riaz, 2000; Akdogan, 1996).

Protein substances undergo structural changes during the extrusion process, whereby their globular structure becomes fibrillar. As a result of denaturation, hydrogen bonds and hydrophobic interactions are disrupted. High temperature and pressure lead to changes in the secondary and tertiary structure of protein substances. At temperatures above 80°C the protein substances denature due to disruption of hydrogen bonds and hydrophobic interactions, and the extrusion temperature is usually above 100°C. Extrusion is a process that changes the structure, solubility and enzymatic digestibility of protein substances (Ivanova, 2018; Penov and Petrova, 2014; Guy, 2001; Riaz, 2000). The effect of extrusion on the nutritional digestibility of proteins and carbohydrates is usually established by biological experiments with animals or by using *in vitro* methods (Ivanova, 2018; Penov and Petrova, 2014; Guy, 2001; Riaz, 2000; Alonso et al., 2000a, b; Ibanoglu et al., 1997; Onyango et al., 2005). During the extrusion of protein-rich foods containing reducing sugars, a Maillard reaction may occur, which would reduce their digestibility (Moss and Otten, 1989; Sgaramella and Ames, 1993).

Starch also undergoes significant changes in the extrusion process. Due to the different structure and functional properties of amylose and amylopectin, they have different behaviors in different treatments (Wang et al., 1993; Ivanova, 2018; Penov and Petrova, 2014). In extrusion, gelatinization is achieved through a combination of moisture, heat, mechanical energy and pressure difference. As a result, several times higher rates of starch degradation are achieved compared to non-extruded foods (Cai et al., 1995; Della Valle et al., 1989; Guy, 2001; Riaz, 2000; Ivanova, 2018; Penov and Petrova, 2014; Alonso et al., 2000a, b; Ibanoglu et al., 1997; Onyango et al., 2005; Colonna et al., 1984; Lue et al., 1991).

The main types of dietary fiber - cellulose, lignin, hemicellulose, pectins, and oligosaccharides - also undergo important changes. Heat treatments can significantly affect the chemical composition and physical properties of dietary fiber (Björck et al., 1984; Ivanova, 2018; Penov and Petrova, 2014). Extrusion leads to an increase in the amount of soluble fibers, due to the mechanical rupture of glycosidic bonds in them (Ivanova, 2018; Penov and Petrova, 2014; Larrea et al., 2005; Lue et al., 1991;

Vasanthan et al., 2002). In other cases, an increase in insoluble fiber has also been observed due to starch gelatinization and the Maillard reaction (Unlu and Faller, 1998; Vasanthan et al., 2002; Esposito et al., 2005).

Vitamins have different resistance to environmental factors (Ivanova, 2018; Penov and Petrova, 2014). The degree of preservation of vitamins during processing usually decreases with increasing temperature and/or residence time of the material in the extruder (Ivanova, 2018; Penov and Petrova, 2014).

The content of unwanted nutrients - trypsin inhibitors, proteases and amylase, lectins, saponins, alkaloids, polysaccharides, phytates and others is significantly reduced during extrusion (Rehman and Shah, 2005; Wang and Daun, 2006; Abd El-Hady and Habiba, 2003; Nwabueze, 2006; Onyango et al., 2005; Poel et al., 1992; Rivas-Vega et al., 2006; Ivanova, 2018; Penov and Petrova, 2014).

The extrusion process also has a significant effect on the antioxidant profile of the products. Some authors consider the changes to be positive and others to be negative. In general, the impact of extrusion on these food indicators is not yet fully understood (Ivanova, 2018; Penov and Petrova, 2014).

4.2.2. Application of Brewers' Spent Greains in Tailor-Made Foods

BSG are used in the development of new types of food products in order to increase their content of dietary fiber. As a result, these novel foods are thought to have preventive effect against diseases such as cancer, diabetes, gastrointestinal disorders and coronary heart disease. The increase in the fiber content is achieved when 30% BSG are included in wheat flour in the production of fortified diet breads (Ivanova, 2018; Penov and Petrova, 2014; Hassona, 1993; Miranda et al., 1994a, b; Ozturk et al., 2002; Mussatto et al., 2006).

30% of the total protein content of BSG is allocated to amino acids and 14.3% to lysine. In addition, leucine, valine, alanine, serine, glycine, glutamic acid, aspartic acid, proline, threonine, cysteine, histidine, isoleucine, methionine, phenyl alanine are also found. BSG provides a significant amount of dietary fiber and polyphenols, which significantly improve food quality. The main amounts of phenolic compounds are phenolic acids, flavanoids, tannins, proanthocyanidins and anophenols. These components have pronounced antioxidant capacity, especially in terms of free radical scavenging. Barley husk is rich in phenolic compounds and when properly mashed they remain in the malt BSG at the end of the process. BSG are suitable for new types of foods due to the presence of biotin, choline, folic acid, niacin, panthenic acid, riboflavin, and thiamine. They contain a significant amount of calcium, cobalt, cuprum, ferrum, magnesium, phosphorus and others. In addition, foods produced with BSG are enriched with triacylglycerols, diacylglycerols, fatty acids, sterols and their esters, glycosides and a significant amount of carbohydrates (Ivanova, 2018; Mussatto et al., 2006; Moreira et al., 2013).

BSG can be directly included in food as a waste product from the brewing industry. The data show that BSG are successfully used in the production of bread and bakery products. BSG flour has a number of advantages: it is easy to make mixtures with; has a high caloric content - 27 MJ/kg; has a high capacity for water adsorption; provides valuable components; weakly adsorbs fat; has a high content of fiber and protein; has a constant color, aroma of the mixture and aroma after roasting. The addition of 10% BSG to bread increases the protein content by 50% and the content of essential amino acids by 10%, the fiber content by 10%, compared to traditional breads without BSG. In addition, BSG breads have 7% fewer calories than traditional ones. The addition of 30% BSG increased the fiber content in pasta to 19.8% compared to control variants. This also provides a prebiotic effect to the food product. BSG are products with a proven prebiotic effect. They provide a synbiotic effect in combination with probiotic bacteria, which makes them suitable for the prevention of diseases of the gastrointestinal tract (Ivanova, 2018; Mussatto et al., 2006; Hassona, 1993; Miranda et al., 1994a, b).

The BSG physicochemical composition makes them suitable for extrusion, but currently there are a very limited number of studies in this direction. However, it was found that the addition of up to 30% BSG in expansion mixtures increases the protein content, phytic acid content, density, reduces the local expansion index and the unit pore area and total pore area of the extrudate. Higher BSG levels affect pores with thinner walls and a rougher surface. The experiments performed by Ainsworth et al. (2007) show varying degrees of changes in the extrudates. For example, an increase in the protein content with 30-50% in extrudates containing BSG is observed. This is due to the higher protein content in BSG and wheat flour. At the same time, the fat content of the extrudates increases, as the increase is directly proportional to the BSG amount. The authors found that the fat content is also affected by the stirrer rate. As BSG contain significant amount of fiber, this has a positive effect on the final product. Increased antioxidant activity, increased content of phenolic compounds and phytic acid are observed. Research data show that the presence of BSG increases protein digestibility due to the extrusion of the product. This is due to the combined influence of shear stresses, heating and pressure. The BSG content also affects the water adsorption index and the water solubility index. These two parameters are used to evaluate the functional characteristics of the products. It was found that the index of water adsorption is influenced by the amount of BSG used, but not by the frequency of rotation of the screw of the extruder. In conclusion, it can be said that the increase in the mechanical energy during extrusion contributes to the increased molecular destruction of the components of BSG flour, which in turn increases the adsorption index of the obtained extrudates. BSG also have positive effect on the starch content of the extrudates. At the same time, the color characteristics of the obtained products change. The data show that the lack of BSG leads to an increase in the brightness of the samples, while a content of 20-30% of BSG increases the redness of the extrudates (Ivanova, 2018; Ainsworth et al., 2007).

ACKNOWLEDGMENTS

The present chapter was supported by the Bulgarian Ministry of Education and Science under the National Research Programme "Healthy Foods for a Strong Bio-Economy and Quality of Life" approved by DCM № 577/17.08.2018.

REFERENCES

Abd El-Hady, E., Habiba, R. (2003). Effect of Soaking and Extrusion Conditions on Antinutrients and Protein Digestibility of Legume Seeds. *Lebensmittel-Wissenschaft und-Technologie*, 36 (3), 285-293. https://doi.org/10.1016/S0023-6438(02)00217-7.

Adeniran, H. A., Abiose, S. H., Ogunsua, A. O. (2008). Production of Fungal β-amylase and Amyloglucosidase on Some Nigerian Agricultural Residues. *Food Bioprocess Technol.*, 3 (5), 693-698.

Ainsworth, P., Ibanoglu, S., Plunkett, A., Ibanoglu, E., Stojceska, V., (2007). Effect of brewers spent grain addition and screw speed on the selected physical and nutritional properties of an extruded snack. *Journal of Food Engineering*, 81, 702–709. https://doi.org/10.1016/j.jfoodeng.2007.01.004.

Akdogan, H. (1996). Pressure, Torque, and Energy Responses of a Twin Screw Extruder at High Moisture Contents. *Food Research International*, 29 (5-6), 423-429.

Albarracin, M., Gonzalez, R. J., Drago, S. R. (2013). Effect of soaking process on nutrient bio-a ccessibility and phytic acid content of brown rice cultivar. *LWT - Food Sci. Technol.*, 53(1), 76-80. https://doi.org/10.1016/j.lwt.2013.01.029.

Aliyu, S., Bala, M. (2011). Brewer`s spent grain. A review of its potentials and applications. *Afr. J. Biotech.* 10 (3), 324-331. https://doi.org/10.5897/AJBx10.006.

Alonso, R., Aguirre, A., Marzo, F. (2000b). Effects of extrusion and traditional processing methods on antinutrients and *in vitro* digestibility of protein and starch in feba and kindey beans. *Food Chemistry*, 68, 159-165. https://doi.org/10.1016/S0308-8146(99) 00169-7.

Alonso, R., Orue, E., Zabalza, M., Grant, G., Marzo, F. (2000a). Effect of extrusion cooking on structure and functional properties of pea and kindey bean proteins. *Journal of the Science of Food and Agriculture*, 80, 397-403. https://doi.org/10.1002/1097-0010 (200002)80:3<397::AID-JSFA542>3.0.CO;2-3.

Antalick, G., Šuklje, K., Blackman, J. W., Meeks, C., Deloire, A., Schmidtke, L. M. (2015). Influence of grape composition on red wine ester profile: comparison between Cabernet sauvignon and Shiraz cultivars from Australian warm climate. *Journal of Agricultural and Food Chemistry*, 63(18), 4664-4672. https://doi.org/10.1021/acs.jafc.5b00966.

Bailey, J. E. Ollis., D. F. (1986). *Biochemical Engineering Fundamentals*, 2nd edition, McGraw-Hill.

Bartolome, B., Garcia-Conesa, M. T., & Williamson, G. (1996). Release of the bioactive compound, ferulic acid, from malt extracts. *Biochemical Society Transactions,* 24, 379S.

Bartolomè, B., Santos, M., Jimenez, J. J., del Nozal, M. J., Gomez-Cordoves C. (2002). Pentoses and hydroxycinnamic acids in brewers' spent grain. *J. Cereal Sci*. 36. 51-58. https://doi.org/10.1006/jcrs. 2002.0442.

Biryukov, V. (2004). *Basics of industrial microbiology.* КолосС, Химия, Moscow. (in Russian).

Björck, I., Asp, N., Dahlqvist, A. (1984). Protein nutritional value of extrusion-cooked wheat flours. *Food Chemistry*, 15 (3), 203-214. https://doi.org/10.1016/0308-8146(84)90004-9.

Boekhout, T., Pfaff, H. (2003). Yeast biodiversity, In: *Yeasts in Food.* Boekhout, T., Robert, V. (Eds.), Behr's Verlag, Hamburg, 7 – 11.

Boulton, C. (2006). Fermentation of beer. In: *Brewing—New Technologies.* Bamforth, C. W. (Ed.), CRC Press, Boca Raton, FL, 228–253.

Boulton, C., Quain, D. (2001). *Brewing yeast and fermentation*, Oxford, Blackwell science.

Branyik, T., Vicente, A., Dostalek, P., Teixeira, J. A. (2005). Continuous beer fermentation using immobilized yeast cell bioreactor systems. *Biotechnol. Prog.* 21, 653-663. https://doi.org/10.1021/ bp050012u.

Briggs, D. E., Hough, J. S., Stevens, R., Young, T. W. (1981). Malt and sweet wort. *Malting and brewing science.* New York: Chapman and Hall. p. 1–387.

Briggs, D. E., Hough, J. S., Stevens, R., Young, T. W. (1981). The biochemistry of malting grain. *Malting and brewing science.* London: Chapman and Hall. 57–109.

Briggs, D., C. Boulton, P. Brookes, R. Stevens (2004). *Brewing science and practice*. Boston, Woodhead publishing in food and science.

Buruk Sahin, Y., Aktar Demirtas, E., Burnak, N. (2016). Mixture design: A review of recent applications in the food industry. *Pamukkale Üniversitesi Mühendislik Bilimleri Dergisi.*, 22(4), 297-304.

Cai, W., Diosady, L., Rubin, L. (1995). Degradation of wheat starch in a twin- screw extruder. *Journal of Food Engineering*, 26 (3), 289- 300. https://doi.org/10.1016/0260-8774(94)00056-F.

Cammerer, B., Jalyschko, W., Kroh, L. W. (2002). Intact carbohydrate structures as part of the melanoidin skeleton. *J Agric Food Chem*, 50(7), 2083–2087. https://doi.org/10.1021/jf011106w.

Carvalho, D. O., Correia, E., Lopes, L., Guido, L. F. (2014). Further insights into the role of melanoidins on the antioxidant potential of barley malt. *Food Chem* 160, 127–133. https://doi.org/10.1016/j.foodchem. 2014.03.074.

Carvalho, D. O., Gonçalves, L. M., Guido, L. F. (2016). Overall antioxidant properties of malt and how they are influenced by the individual constituents of barley and the malting process. *Comprehensive Reviews in Food Science and Food Safety*, 15, 927-943. https://doi.org/10.1111/ 1541-4337.12218.

Cechovska, L., Konecny, M., Velısek, J., Cejpek, K. (2012). Effect of maillard reaction on reducing power of malts and beers. *Czech J Food Sci*, 30(6), 548–556.

Chandra, C. S., Buggey, L. A., Peters, S., Cann, C., Liegeois, C. (2001). Factors affecting the development of antioxidant properties of malts during the malting and roasting process. *Project Report no. 242*. B. R. International.

Charalampopoulos, D., Pandiella, S. S., Webb, C. (2003). Evaluation of the effect of malt, wheat and barley extracts on the viability of potentially probiotic lactic acid bacteria under acidic conditions. *International Journal of Food Microbiology*, 82(2), 133-141. https://doi.org/10.1016/ S0168-1605(02)00248-9.

Charalampopoulos, D., Wang, R., Pandiella, S. S., Webb, C. (2002). Application of cereals and cereal components in functional foods: a review. *Int J Food Microbiol.*, 79(1-2), 131-141. https://doi.org/10.1016/s0168-1605(02)00187-3.

Coghe, S., Adriaenssens, B., Leonard, S., Delvaux, F. R. (2004). Fractionation of colored Maillard reaction products from dark specialty malts. *J Am Soc Brew Chemist*, 62(2), 79–86. https://doi.org/ 10.1094/ASBCJ-62-0079.

Coghe, S., D'Hollander, H., Verachtert, H., Delvaux, F. R. (2005). Impact of dark specialty malts on extract composition and wort fermentation. *J Inst Brew*, 111(1), 51–60. https://doi.org/10.1002/j.2050-0416.2005.tb00648.x.

Coghe, S., Gheeraert, B., Michiels, A., Delvaux, F. R. (2006). Development of Maillard reaction related characteristics during malt roasting. *J Inst Brew* 112(2), 148–156. https://doi.org/10.1002/j.2050-0416.2006.tb00244.x.

Colonna, P., Mercier, C. (1983). Macromolecular modifications of manioc starch components by extrusion - cooking with and without lipids. *Carbohydrate Polymers*, 3, 87-108. https://doi.org/10.1016/0144-8617(83)90001-2.

Cortes, N., Kunz, T., Suarez, A. F., Hughes, P., Methner, F. J. (2010). Development and correlation between the organic radical concentration in different malt types and oxidative beer stability. *J Am Soc Brew Chemist*, 68(2), 107–113. https://doi.org/10.1094/ASBCJ-2010-0412-01.

De Keukeleire, D. (2000). Fundamentals of beer and hop chemistry. *Quimica Nova*, 23, 108–112. https://doi.org/10.1590/S0100-40422 000000100019.

Debourg, A., L. van Nedervelde (1999) The use of dried yeast in the brewing industry, In: *Proceedings of the 27th Congress of the European Brewery Convention*, Cannes, 751-760.

Dvorakova, M., Guido, L. F., Dostalek, P., Skulilova, Z., Moreira, M. M., Barros, A. A. (2008a). Antioxidant properties of free, soluble ester and insoluble-bound phenolic compounds in different barley varieties and corresponding malts. *J Inst Brew*, 114(1), 27–33. https://doi.org/ 10.1002/j.2050-0416.2008.tb00302.x.

Dvorakova, M., Moreira, M. M., Dostalek, P., Skulilova, Z., Guido, L. F., Barros, A. A. (2008c). Characterization of monomeric and oligomeric flavan-3-ols from barley and malt by liquid chromatography-ultraviolet detection-electrospray ionization mass spectrometry. *J Chromatogr A*, 1189(1–2), 398–405. https://doi.org/10.1016/j.chroma.2007.10.080.

Eaton, B. (2006). An overview of brewing, In: *Handbook of brewing*, Priest, D., Stewart, G. (Edt.), Tylor and Francis group, 77-913.

Esposito, F., Arlotti, G., Bonifati, A., Napolitano, A., Vitale, D., Fogliano, V. (2005). Antioxidant activity and dietary fibre in durum wheat bran by-products. *Food Research International*, 38 (10), 1167-1173. https://doi.org/10.1016/j.foodres.2005.05.002.

Fulgencio, S.-C., Serranom, J., Pérez-Jiménez, J. (2009). What Contribution Is Beer to the Intake of Antioxidants in the Diet?, In: *Beer in Health and Disease Prevention*, Preedy, V. (Ed.), Academic Press, 441-448. https://doi.org/10.1016/B978-0-12-373891-2.00042-0.

Gerhäuser, C. (2005). Beer constituents as potential cancer chemopreventive agents. *Eur. J. Cancer*, 41, 1941–1954. https://doi.org/10.1016/j.ejca.2005.04.012.

Ghiselli, A., Natella, F., Guidi, A., Montanari, L., Fantozzi, P., Scaccini, C. (2000). Beer increases plasma antioxidant capacity in humans. *J. Nutr. Biochem.*, 11, 76 – 80. https://doi.org/10.1016/S0955-2863(99)00077-7.

Goupy, P., Hugues, M., Boivin, P., Amiot, M. J. (1999). Antioxidant composition and activity of barley (Hordeum vulgare) and malt extracts and of isolated phenolic compounds. *J Sci Food Agric*, 79(12), 1625–1634. https://doi.org/10.1002/(SICI)1097-0010(199909)79:12<1625::AID-JSFA411>3.0.CO;2-8.

Greiner, R., Konietzny, U. (2006). Phytase for food applications. *Food Technol. Biotechnol.*, 44, 125-140.

Guy, R. (2001). *Extrusion Cooking, Technologies and Application*. Florida. Woodhead Publication. p. 151.

Hassani, A., Procopio, S., Becker T. (2016). Influence of malting and lactic acid fermentation on functional bioactive components in cereal-based raw materials: a review paper. *International Journal of Food Science & Technology*, 51, 14–22. https://doi.org/10.1111/ijfs.12965.

Hassani, A., Zarnkow, M., Becker, T. (2013). Influence of malting conditions on sorghum (*Sorghum bicolor* (L.) Moench) as a raw material for fermented beverages. *Food Science and Technology International*, 20, 453–463. https://doi.org/10.1177/1082013213490710.

Hassona, H. Z., 1993. High fibre bread containing brewer's spent grains and its effect on lipid metabolism in rats. *Die Nahrung* 37, 576–582. https://doi.org/10.1002/food.19930370609.

Hernanz, D., Nuñez, V., Sancho, A. I., Faulds, C. B., Williamson, G., Bartolomé, B, Gómez-Cordovés, C. (2001). Hydroxycinnamic acids and ferulic acid dehydrodimers in barley and processed barley. *J. Agric. Food Chem.* 49, 4884–4888. https://doi.org/10.1021/ jf010530u.

Hiroyasu Tobe, H. (2009). Biological activities of humulone, In: *Beer in Health and Disease Prevention*, Preedy, V. (Ed.), Academic Press, 695-702. https://doi.org/10.1016/B978-0-12-373891-2.00070-5.

Holtekjolen, A. K., Kinitz, C., Knutsen, S. H. (2006). Flavanol and bound phenolic acid contents in different barley varieties. *J Agric Food Chem*, 54(6), 2253–2260. https://doi.org/10.1021/jf052394p.

Huige, N. J. (1994) Brewery by-products and effluents, in *Handbook of Brewing*. Hardwick, W. A. (Ed.), pp. 501–550, Marcel Dekker, New York.

Ibanoglu S., P. Ainsworth, G. Hayes (1997). *In vitro* protein digestibility and content of thiamin and riboflavin in extruded tarhana, a traditional Turkish cereal food. *Food Chemistry*, 58 (1-2), 141-144. https://doi.org/10.1016/S0308-8146(96)00233-6.

Inns, E. L., Buggey, L. A., Booer, C., Nursten, H. E., Ames, J. M. (2007). Effect of heat treatment on the antioxidant activity, color, and free phenolic acid profile of malt. *J Agric Food Chem*, 55(16), 6539–6546. https://doi.org/10.1021/jf0710231.

Inns, E. L., Buggey, L. A., Booer, C., Nursten, H. E., Ames, J. M. (2011). Effect of modification of the kilning regimen on levels of free ferulic acid and antioxidant activity in malt. *J Agric Food Chem*, 59(17), 9335–9343. https://doi.org/10.1021/jf201253t.

Ivanov, K., Petelkov, I., Shopska, V., Denkova, R., Gochev, V., Kostov, G. (2016). Investigation of mashing regimes for low-alcohol beer production. *Journal of the Institute of Brewing*, 122(3), 508-516. https://doi.org/10.1002/jib.351.

Ivanova, K., (2018). *Technological and functional properties of extrudates enriched with brewers' spent grains*, PhD Thesis, Plovdiv. (in Bulgarian).

Ivanova, K., Denkova, R., Kostov, G., Petrova, T., Bakalov, I., Ruscova, M., and Penov, N. (2017). Extrusion of brewers' spent grains and application in the production of functional food. Characteristics of spent grains and optimization of extrusion. *J. Inst. Brew.*, 123, 544– 552. https://doi.org/10.1002/jib.448.

Jurkova, M., Horak, T., Haskova, D., Culık, J., Cejka, P., & Kellner, V. (2012). Control of antioxidant beer activity by the mashing process. *Journal of the Institute of Brewing*, 118(2), 230–235. https://doi.org/10.1002/jib.35.

Kabzev, Y., I. Ignatov. (2011). *Technology of beer*, 1st edition, University of Food Technologies-Plovdiv Academic Press, p. 388 (in Bulgarian).

Khidzir, K. M., Noorlidah, A., Agamuthu, P. (2010). Brewery spent grain. chemical characteristics and utilization as an enzyme substrate. *Malaysian J. Sci.* 29(1), 41-51. https://doi.org/10.22452/mjs.vol29 no 1.7.

Kostov, G. (2015). *Intensification of fermentation processes with immobilized biocatalisists*. DSc Thesis, p. 307.

Krahl, M. (2010). Dissertation, WZW TU München.

Kreisz, S. (2009). Malting. In: *Handbook of Brewing: Processes, Technology, Markets.* Eslinger H. M. (ed.), Wiley-VCH Verlag GmbH & Co. KGaA, Weinheim, 147-164. https://doi.org/10.1002/9783527 623488.ch6.

Kunze, W. (2004). *Technology of brewing and malting*, 3rd international edition, VLB, Berlin.

Landete, J. M. (2013). Dietary intake of natural antioxidants: vitamins and polyphenols. *Crit Rev Food Sci Nutr*, 53(7), 706–721. https://doi.org/ 10.1080/10408398.2011.555018.

Larrea, M., Chang, Y., Bustos, F. (2005). Effect of some operational extrusion parameters on the constituents of orange pulp. *Food*

Chemistry, 89 (2), 301-308. https://doi.org/10.1016/j.foodchem.2004.02.037.

Leitao, C., Marchioni, E., Bergaentzle, M., Zhao, M., Didierjean, L., Miesch, L., Holder, E., Miesch, M., Ennahar, S. (2012). Fate of polyphenols and antioxidant activity of barley throughout malting and brewing. *J Cereal Sci* 55(3), 318–322. https://doi.org/10.1016/j.jcs.2012.01.002.

Leitao, C., Marchioni, E., Bergaentzle, M., Zhao, M., Didierjean, L., Taidi, B., Ennahar, S. (2011). Effects of processing steps on the phenolic content and antioxidant activity of beer. *J Agric Food Chem*, 59(4), 1249–1255. https://doi.org/10.1021/jf104094c.

Lengyel, E., Panaitescu, M. (2017). The management of selected yeast strains in quantifying terpene flavours in wine, *Management of Sustainable Development, Sibiu, Romania*, 9 (1), 27-30. https://doi.org/10.1515/msd-2017-0010.

Lu, J., Zhao, H., Chen, J., Fan, W., Dong, J., Kong, W., Sun, J., Cao, Y., Cai, G. (2007). Evolution of phenolic compounds and antioxidant activity during malting. *J Agric Food Chem*, 55(26), 10994–11001. https://doi.org/10.1021/jf0722710.

Lue, S., Hsieh, F., Huff, H. (1991). Extrusion cooking of corn meal and sugar beet fiber. effects on expansion properties, starch gelatinization, and dietary fiber content. *Cereal Chemistry*, 68, 227-234.

Madhujith, T., Izydorczyk, M., Shahidi, F. (2006). Antioxidant properties of pearled barley fractions. *J Agric Food Chem*, 54(9), 3283–3289. https://doi.org/10.1021/jf0527504.

Magalhaes, P. J., Almeida, S. M., Carvalho, A. M., Goncalves, L. M., Pacheco, J. G., Cruz, J. M., Guido, L. F., Barros, A. A. (2011). Influence of malt on the xanthohumol and isoxanthohumol behavior in pale and dark beers: a micro-scale approach. *Food Res Intl*, 44(1), 351–359. https://doi.org/10.1016/j.foodres.2010.10.008.

Maillard, M. N., & Berset, C. (1995). Evolution of antioxidant activity during kilning: Role of insoluble bound phenolic acids of barley and malt. *Journal of Agricultural and Food Chemistry*, 43, 1789–1793. https://doi.org/10.1021/jf00055a008.

McCarthy, A. L., O'Callaghan, Y. C., Connolly, A., Piggott, C., FitzGerald, R. J., O'Brien, N. (2012) Phenolic extracts of brewers' spent grain (BSG) as functional ingredients — Assessment of their DNA protective effect against oxidant-induced DNA single strand breaks in U937 cells. *Food Chem. 134 (2)*, 641–646. https://doi.org/10.1016/j.foodchem. 2012.02.133.

Mendes, I., Sanchez, I., Franco-Duarte, R., Camarasa, C., Schuller, D., Dequin, S., Sousa, M.(2017). Integrating transcriptomics and metabolomics for the analysis of the aroma profiles of *Saccharomyces cerevisiae* strains from different origin. *BMC Genomics*, 18, 455-464. https://doi.org/10.1186/s12864-017-3816-1.

Meussdoerffer, F., Zarnkow, M. (2009). Starchy Raw Materials, In: *Handbook of Brewing: Processes, Technology, Markets.* Eslinger H. M. (ed.), Wiley-VCH Verlag GmbH & Co. KGaA, Weinheim, 43-83. https://doi.org/10.1002/9783527623488.ch2.

Miranda, M. Z., Grossmann, M. V. E., Nabeshima, E. H. (1994a). Utilization of brewers' spent grain for the production of snacks with fiber. 1. Physicochemical characteristics. *Brazilian Archives of Biology and Technology*, 37, 483–493.

Miranda, M. Z., Grossmann, M. V. E., Nabeshima, E. H. (1994b). Utilization of brewer spent grain (BSG) for production of snacks with fiber. 2. Sensory analysis of snacks. *Brazilian Archives of Biology and Technology*, 37, 9–21.

Morales, F. J, Fernandez-Fraguas, C., Jimenez-Perez, S. (2005). Iron-binding ability of melanoidins from food and model systems. *Food Chem*, 90(4), 821–827. https://doi.org/10.1016/j.foodchem. 2004.05. 030.

Moreira, M. M., Simone Morais, S., Carvalho, D. O., Barros, A. A., Delerue-Matos, C., Guido, L. F. (2013). Brewer's spent grain from different types of malt. Evaluation of the antioxidant activity and identification of the major phenolic compounds. *Food Res. Inter. 54*, 382–388. https://doi.org/10.1016/j.foodres.2013.07.023.

Moss, J., Otten, L. (1989). A relationship between color development and moisture content during roasting of peanut. *Canadian Institute of Food*

Science and Technology Journal, 22, 34-39. https://doi.org/10.1016/S0315-5463(89)70298-4.

Munroe, J. (2006). Fermentation, In: *Handbook of brewing, Priest*, D., G. Stewart (Eds.), Tylor and Francis group, 487-524.

Mussato, S., Dragone, G., Roberto, I. C. (2006). Brewers' spent grain. generation, characteristics and potential applications. *J. Cereal Sci.* 43, 1–14. https://doi.org/10.1016/j.jcs.2005.06.001.

Mussato, S., Roberto, I. C. (2006). Chemical characterization and liberation of pentose sugars from brewer's spent grain. *J. Chem. Technol. Biotechnol.* 81. 268-274. https://doi.org/10.1002/jctb.1374.

Mussatto, S. I., Rocha, G., Roberto, I. C. (2008). Hydrogen peroxide bleaching of cellulose pulps obtained from brewer's spent grain. *Cellulose* 15.641-649.

Nardini, M., Ghiselli, A. (2004). Determination of free and bound phenolic acids in beer. *Food Chem.*, 84, 137–143. https://doi.org/10.1016/S0308-8146(03)00257-7.

Narziss, L. (1999). *Die Bierbrauerei: Band In: Die Technologie der Malzbereitung* [*The brewery: Volume In: The technology of malt preparation*], 7th edition, Ferdinand Enke Verlag, Stuttgart, 21 – 65.

Nordkvist, E., Salomonsson, A. C., Aman, P. (1984). Distribution of insoluble bound phenolic acids in barley grain, *J. Sci. Food Agric.* 35, 657 – 661. https://doi.org/10.1002/jsfa.2740350611.

Nwabueze, T. (2006). Gelatinization and viscosity behavior of single – screw extruded african breadfruit (*Treculia Africana*) mixtures. *Journal of Food Processing and Preservation*, 30, 717-731. https://doi.org/10.1111/j.1745-4549.2006.00100.x.

Onyango, C., Noetzold, H., Ziems, A., Hofmann, T., Bley, T., Henle, T. (2005). Digestibility and antinutrient properties of acidified and extruded maize – finger millet blend in the production of *Uji*. *LWT - Food Science and Technology*, 38 (7), 697-707. https://doi.org/10.1016/j.lwt.2004.09.010.

Öztürk, S., Özboy, O., Cavidoglu, I., Köksel, H. (2002). Effects of brewers' spent grains on the qualityand dietary fibre content of cookies. *J. Inst.*

Brew. 108(1), 23-27. https://doi.org/10.1002/j.2050-0416.2002.tb00116.x.

Penov, N., Petrova, T. (2014). *Extrusion - a technology for food production*, University of Food Technologies Academic Press, ISBN 978-954-24-0255-8, p. 248. (in Bulgarian).

Poel. A., Stolp, W., Zuilichem, D. (1992). Twin-screw extrusion of two pea varieties. Effects of temperature and moisture level on antinutritional factors and protein dispersibility. *Journal of the Science of Food and Agriculture*, 58 (1), 83-87. https://doi.org/ 10.1002/jsfa.2740580114.

Polak, J., Bartoszek, M., Stanimirova, I. (2013). A study of the antioxidant properties of beers using electron paramagnetic resonance. *Food Chem*, 141(3), 3042–3049. https://doi.org/10.1016/j.foodchem.2013.05.133.

Qingming, Y., Xianhui, P., Weibao, K., Hong, Y., Yidan, S., Li, Z., Yanan, Z., Yuling, Y., Lan, D., Guoan, L. (2010). Antioxidant activities of malt extract from barley (Hordeum vulgare L.) toward various oxidative stress in vitro and in vivo. *Food Chem,* 118(1), 84–89. https://doi.org/ 10.1016/j.foodchem.2009.04.094.

Quifer-Rada, P., Vallverdu-Queralt, A., Martınez-Huelamo, M., Chiva-Blanch, G., Jauregui, O., Estruch, R., Lamuela-Raventos, R. (2015). Overall antioxidant properties of malt characterisation of beer polyphenols by high resolution mass spectrometry (LC-ESI-LTQ-Orbitrap-MS). *Food Chem*, 169, 336–343. https://doi.org/10.1016/ j.foodchem.2014.07.154.

Rao, R. S., Muralikrishna, G. (2006). Water soluble feruloyl arabinoxylans from rice and ragi: changes upon malting and their consequence on antioxidant activity. *Phytochemistry*, 67(1). 91-99. https://doi.org/ 10.1016/j.phytochem.2005.09.036.

Rehman, Z., Shah, W. (2005). Thermal heat processing effects on antinutrients, protein and starch digestibility of food legumes. *Food Chemistry*, 91 (2), 327-331. https://doi.org/10.1016/j.foodchem.2004.06.019.

Riaz, M. (2000). Introduction to Extruders and their Principles. *In.* Extruders in Food Applications, Riaz, M. N. (eds.), Boca Raton, Florida. CRC Press, 1-23.

Rivas-Vega, M., Goytortua-Bores, E., Ezquerra-Brauer, J., Salazar-Garcia, M., Cruz-Suarez, L., Nolasco, H., Civera-Cerecedo, R. (2006). Nutritional value of cowpea (*Vigna unguiculata* L. Walp) meals as ingredients in diets for pacific white shrimp (*Litopenaeus vannamei* Boone). *Food Chemistry*, 97 (1), 41-49. https://doi.org/10.1016/j.foodchem.2005.03.021.

Rivero, D., Perez-Magarino, S., Gonzalez-Sanjose, M. L., Valls-Belles, V., Codoner, P., Muniz, P. (2005). Inhibition of induced DNA oxidative damage by beers: correlation with the content of polyphenols and melanoidins. *J Agric Food Chem*, 53(9), 3637–3642. https://doi.org/10.1021/jf048146v.

Rusell I. (2006). Yeast, In: *Handbook of brewing*, Priest, D., G. Stewart (Eds.), Tylor and Francis group, 281-333.

Russ, W., Mortel, H., Meyer-Pittroff, R. (2005). Application of spent grains to increase porosity in bricks. *Constr. Build. Mater. 19*, 117–126. https://doi.org/10.1016/j.conbuildmat.2004.05.014.

Salomonsson, A. C., Theander, O. Aman, P. (1980). Composition of normal and high-lysine barley. *Swed. J. Agric. Res.*, 10, 11–16.

Samaras, T. S., Camburn, P. A., Chandra, S. X., Gordon, M. H, Ames, J. M. (2005). Antioxidant properties of kilned and roasted malts. *J Agric Food Chem*, 53(20), 8068–8074. https://doi.org/10.1021/ jf051410f.

Sánchez-Moreno, C., Larrauri, J. A., Saura-Calixto, F. (1998). A procedure to measure the antiradical efficiency of polyphenols. *J. Sci. Food Agric.* 76, 270–276. https://doi.org/10.1002/(SICI)1097-0010(199802)76:2<270::AID-JSFA945>3.0.CO;2-9.

Santos, M., Jimenez, J., Bartolome, B., Gomez-Cordoves, C., del Nozal, M. J. (2003). Variability of brewers' spent grain within a brewery. *Food Chem. 80*, 17–21. https://doi.org/10.1016/S0308-8146(02)00229-7.

Serafini, M. (2006). The role of antioxidants in disease prevention. *Medicine*, 34, 533–535. https://doi.org/10.1053/j.mpmed.2006.09.007.

Sgaramella, S., Ames, J. (1993). The development and control of colour in extrusion cooked foods. *Food Chemistry*, 46 (2), 129-132. https://doi.org/10.1016/0308-8146(93)90024-A.

Shahidi, F., Naczk, M. (2004). *Phenolics in Foods and Nutraceuticals*. CRC Press LLC, Boca Raton, FL.

Shopska, V., Denkova, R., Kostov, G. (2016). Beer production with encapsulated yeast cells, In: *Beer: Production, Consumption and Health Effects*. Salazar, W. (Ed.), Nova Science Publishers, Inc. Hauppauge, NY, 27 – 100.

Siro, I., E. Kapolna, and A. Lugasi. 2008. "Functional food. Product development, marketing and consumer acceptance – A review." *Journal of Appetite* 51:456 – 467. https://doi.org/10.1016/j.appet. 2008.05.060.

Soccol, C. R., Prado, M. R. M, Garcia, L. M. B, Rodrigues, C., Medeiros, A. B. P, et al. (2014). Current Developments in Probiotics. *J Microb Biochem Technol*, 7, 11-20. https://doi.org/10.4172/1948-5948. 1000175.

Subba Rao, M. V., Muralikrishna, G. (2002). Evaluation of the antioxidant properties of free and bound phenolic acids from native and malted finger millet (ragi, Eleusine coracana Indaf-15). *J Agric Food Chem*, 50(4), 889–892. https://doi.org/10.1021/jf011210d.

Szwajgier, D. (2011). Dry and wet milling of malt. A preliminary study comparing fermentable sugar, total protein, total phenolics and the ferulic acid content in non-hopped worts. *Journal of the Institute of Brewing*, 117(4), 569–577. https://doi.org/10.1002/j.2050-0416. 2011.tb00505.x.

Tange, C. (2009). Yeast, In: *Handbook of Brewing: Processes, Technology, Markets*. Eslinger H. M. (ed.), Wiley-VCH Verlag GmbH & Co. KGaA, Weinheim, 119-145. https://doi.org/10.1002/9783527623488.ch5.

Unlu, E., Faller, F. (2002). RTD in twin-screw food extrusion. *Journal of Food Engineering*, 53, 115-131. https://doi.org/10.1016/S0260-8774 (01)00148-0.

Vanbeneden, N., Gils, F., Delvaux, F., & Delvaux, F. R. (2007). Variability in the release of free and bound hydroxycinnamic acids from diverse malted barley (Hordeum vulgare L.) cultivars during wort production. *Journal of Agricultural and Food Chemistry*, 55, 11002–11010. https://doi.org/10.1021/jf072340a.

Vanbeneden, N., Van Roey, T., Willems, F., Delvaux, F., & Delvaux, F. R. (2008b). Release of phenolic flavour precursors during wort production: Influence of process parameters and grist composition on ferulic acid release during brewing. *Food Chemistry*, 111(1), 83–91. https://doi.org/10.1016/j.foodchem.2008.03.029.

Vanderhaegen, B., Neven, H., Verachtert, H., Derdelinckx, G. (2006). The chemistry of beer aging—a critical review. *Food Chem*, 95(3), 357–381. https://doi.org/10.1016/j.foodchem.2005.01.006.

Vasanthan, T., Gaosong, J., Yeung, J., Li, J. (2002). Dietary fibre profile of barley flour as affected by extrusion cooking. *Food Chemistry*, 77, 35-40. https://doi.org/10.1016/S0308-8146(01)00318-1.

Villaño, D., Fernández-Pachón, S., Troncoso, A. M., García-Parrilla, M. C. (2005). Comparison of antioxidant activity of wine phenolic compounds and metabolites in vitro. *Anal. Chim. Acta*, 538, 391–398. https://doi.org/10.1016/j.aca.2005.02.016.

Virkajarvi, I. (2001). *Feasibility of continuous main fermentation of beer using immobilized yeast*. PhD thesis, Helsinki University of Technology, Espoo.

Wang, H. Y., Qian, H., Yao, W. R. (2011). Melanoidins produced by the Maillard reaction: Structure and biological activity. *Food Chem*, 128(3), 573–584. https://doi.org/10.1016/j.foodchem.2011.03.075.

Wang, N., Daun, J. (2006). Effects of variety and crude protein content on nutrients and anti-nutrients in lentils (*Lens culinaris*). *Food Chemistry*, 95 (3), 493-502. https://doi.org/10.1016/j.foodchem.2005. 02.001

Wang, Y., White, P., Pollak, L., Jane, J. (1993). Characterization of Starch Structures of 17 Maize Endosperm Mutant Genotypes with Oh43 Inbred Line Background. *Cereal Chemistry*, 70 (2), 171-179.

Woodside, J. V., McCall, D., McGartland, C., Young, I. S. (2005). Micronutrients: dietary intake v. supplement use. *Proc. Nutr. Soc.*, 64, 543–553. https://doi.org/10.1079/PNS2005464.

Yahya, H., Linforth, R. S. T, Cook, D. J. (2014). Flavour generation during commercial barley and malt roasting operations: a time course study. *Food Chem*, 145, 378–387. https://doi.org/10.1016/j.foodchem. 2013.08.046.

Yu, J., Vasanthan, T., Temelli, F. (2001). Analysis of Phenolic Acids in Barley by High-Performance Liquid Chromatography, *J. Agric. Food Chem.*, 49, 4352 – 4358. https://doi.org/10.1021/jf0013407.

Zhao, H., Chen, W., Lu, J., Zhao, M. (2010). Phenolic profiles and antioxidant activities of commercial beers. *Food Chem*, 119(3), 1150–1158. https://doi.org/10.1016/j.foodchem.2009.08.028.

BIOGRAPHICAL SKETCHES

Vesela Nevelinova Shopska

Affiliation: Department of Technology of wine and brewing, University of Food Technologies, Plovdiv, Bulgaria

Education:

2001 – 2005 – Bachelor; Beverages technologies; University of Food Technologies

2005 – 2006 – Master; Technology of wine and brewing; University of Food Technologies;

2011 – 2014 - PhD; 5.12 Food technologies (Technology of alcoholic and non-alcoholic beverages); University of Food Technologies

Business Address: 26 Maritza Blvd, Plovdiv, Bulgaria

Research and Professional Experience: Beer production with immobilized cells; Modelling of beer fermentation; Beer microbiology; Antioxidant activity of malt and wort; Functional wort-based non-alcoholic beverages.

Professional Appointments: Chief Assistant Professor

Honors: PhD

Publications from the Last 3 Years:

1) Kostov G., I. Petelkov, R. Denkova, V. Shopska, Z. Denkova, B. Goranov, V. Iliev (2017). Modeling Of Continuous Ethanol Fermentation In Ideal Mixing Column Bioreactor. *Proceedings 31st European Conference on Modelling and Simulation ©ECMS*, pp. 335-341.
2) Kostov G., R. Denkova-Kostova, V. Shopska, P. Nedyalkov, Z. Denkova, B. Goranov, V. Iliev, K. Ivanova, D. Teneva (2018). New approach to modelling the kinetics of the fermentation process in cultivation of lactic acid bacteria. *Proceedings 32nd European Conference on Modelling and Simulation ©ECMS*, pp. 212-218.
3) Lyubenova V., M. Ignatova, G. Kostov, V. Shopska, E. Petre, M. Roman (2018) An Interactive Teaching System for Kinetics Modelling of Biotechnological Processes, *Proceedings of 22nd International Conference on System Theory, Control and Computing (ICSTCC)*, 10-12 October 2018, Sinaia, Romania, 366-371 doi: 10.1109/ICSTCC.2018.8540688.
4) Nedyalkov P., R. Denkova, D. Teneva, V. Shopska, B. Goranov, Z. Denkova, G. Kostov, M. Kaneva (2019) Yeast selection for non-alcoholic and low-alcoholic beverages based on wort. *Food Science and Applied Biotechnology*, 2019, 2(2), 140-148 https://doi.org/10.30721/fsab2019.v2.i2.
5) Kostov G., V. Shopska, R. Denkova-Kostova, B. Goranov (2019). Analytical approaches to determine the specific biomass growth rate in brewing. *Proceedings 33rd European Conference on Modelling and Simulation ©ECMS*, pp. 125-131.
6) Shopska V., R. Denkova, V. Lyubenova, G. Kostov (2019) Kinetic Characteristics of Alcohol Fermentation in Brewing: State of Art and Control of the Fermentation Process. In: *Fermented Beverages Volume 5: The Science of Beverages* (Edited by A. M. Grumezescu, A. M. Holban), Woodhead Publishing, 529-576 https://doi.org/10.1016/B978-0-12-815271-3.00001-4 ISBN: 978-0-12-815271-3 (print) ISBN: 978-0-12-815703-9 (online).

7) Shopska V, M. Dzivoderova-Zarcheva, K. Ivanova, P. Nedyalkov, R. Denkova-Kostova, M. Kaneva, G. Kostov (2019) Phytochemical characteristics of different malts and possibilities for their application in functional beverages. *Proceedings of the X International Scientific Agricultural Symposium "Agrosym 2019,"* 03-06 October 2019, Jahorina, Bosnia and Herzegovina, 785-790 ISBN 978-99976-787-2-0.
8) Kostov, G., V. Iliev, B. Goranov, R. Denkova, V. Shopska (2020) Immobilized cell bioreactors in fermented beverage production: design and modelling, In: 19. *Biotechnological progress and beverage consumption,* The Science of Beverages series (A. M. Grumezescu and A. M. Holban eds.), 339-376. https://doi.org/10.1016/B978-0-12-816678-9.00011-4.
9) Petelkov, I., V. Shopska, R. Denkova-Kostova, G. Kostov, V. Lyubenova, (2020). Investigation of Different Regimes of Beer Fermentation with Free and Immobilized Cells. *Periodica Polytechnica Chemical Engineering.* 64, 2(2020), 162-171. https://doi.org/10.3311/PPch.13845.

Rositsa Stefanova Denkova-Kostova

Affiliation: Department of Biochemistry and molecular biology, University of Food Technologies, Plovdiv, Bulgaria

Education:
2005 – 2009 – Bachelor; Biotechnologies; Sofia University "St. Kliment Ohridski";
2009 – 2011 – Master; Biotechnologies; Sofia University "St. Kliment Ohridski";
2011 – 2014 - PhD; 5.11 Biotechnologies (Technology of bioactive substances); Sofia University "St. Kliment Ohridski."

Business Address: 26 Maritza Blvd, Plovdiv, Bulgaria

Research and Professional Experience:

Isolation of lactic acid bacteria from different sources; Identification of microorganisms using physic-chemical and molecular-genetic methods; Examination of probiotic properties of newly isolated lactic acid bacteria strains; Selection of probiotic strains for development of starters for functional foods and beverages; Antimicrobial activity of potentially probiotic strains, essential oils, plant extracts against pathogenic and saprophytic microorganisms.

Professional Appointments: Chief Assistant Professor

Honors: PhD

Publications from the Last 3 Years:

1) Denkova, R., B. Goranov, Z. Denkova, D. Teneva, G. Kostov (2017). Enhancing yogurt health benefits: Development of starters for dairy and non-dairy yogurt. *Functional Foods: Sources, Health Effects and Future Perspectives.* Nova Science Publishers, Inc. Hauppauge, NY, 221 - 282.
2) Denkova, R., G. Kostov, Z. Denkova (2017). Functional bread: development of sourdough starters to improve bread quality. *Functional Foods: Sources, Health Effects and Future Perspectives.* Nova Science Publishers, Inc. Hauppauge, NY, 43 - 120.
3) Kostov G., I. Petelkov, R. Denkova, V. Shopska, Z. Denkova, B. Goranov, V. Iliev (2017). Modeling Of Continuous Ethanol Fermentation In Ideal Mixing Column Bioreactor. *Proceedings 31st European Conference on Modelling and Simulation ©ECMS,* pp. 335-341.
4) Teneva D., R. Denkova, B. Goranov, Z. Denkova, G. Kostov (2017). Antimicrobial activity of Lactobacillus plantarum strains against Salmonella pathogens. *Ukrainian Food Journal. 2017. Volume 6. Issue 1: 125 – 133.*
5) Denkova R., B. Goranov, D. Teneva, Z. Denkova, G. Kostov (2017). Antimicrobial activity of probiotic microorganisms:

mechanisms of interaction and methods of examination. In: *Antimicrobial Research: Novel bioknowledge and educational programs"* (Microbiology Book Series - Volume #6) 201 - 212.
6) Ivanova K., R. Denkova, G. Kostov, T. Petrova, I. Bakalov, M. Ruscova, N. Penov (2017). Extrusion of brewers' spent grains and application in the production of functional food. Characteristics of spent grains and optimization of extrusion. *J. Inst. Brew.,* 123 (4), pp. 544-552. doi 10.1002/jib.448.
7) Ahmedova G., M. Kaneva, R. Denkova, D. Teneva, Z. Denkova, I. Ignatov, P. Nedyalkov (2017). Dynamics of the fermentation process of a lactic acid beverage based on wort and mint with Lactobacillus casei ssp. rhamnosus Oly. *Scientific Works of the University of Food Technologies 64(1).*
8) Kostov G., R. Denkova-Kostova, V. Shopska, P. Nedyalkov, Z. Denkova, B. Goranov, V. Iliev, K. Ivanova, D. Teneva (2018). New approach to modelling the kinetics of the fermentation process in cultivation of lactic acid bacteria. *Proceedings 32nd European Conference on Modelling and Simulation ©ECMS,* pp. 212-218.
9) Vasileva I., R. Denkova, R. Chochkov, D. Teneva, Z. Denkova, Tz. Dessev, P. Denev, A. Slavov (2018). Effect of lavender *(Lavandula angustifolia)* and melissa (*Melissa Officinalis*) waste on quality and shelf life of bread. *Food Chemistry,* 253(1): 13-21.
10) Denkova-Kostova R., B. Goranov, D. Teneva, Z. Denkova, G. Kostov (2018). Antimicrobial activity of *Lactobacillus* strains against *Escherichia coli*: a multimethod approach to explore the mechanisms and factors determining the antimicrobial action. In: Antimicrobial Research: Novel bioknowledge and educational programs" (Microbiology Book Series - Volume #7) (Enrique Torres-Hergueta and A. Méndez-Vilas Eds.), *Microbiology Book Series,* 7, 43 – 54, ISBN: 978-84-947512-5-7.
11) Nedyalkov P., R. Denkova, D. Teneva, V. Shopska, B. Goranov, Z. Denkova, G. Kostov, M. Kaneva (2019) Yeast selection for non-alcoholic and low-alcoholic beverages based on wort. *Food Science*

and *Applied Biotechnology,* 2019, 2(2), 140-148 https://doi.org/10.30721/fsab2019.v2.i2.

12) Teneva D., R. Denkova-Kostova, B. Goranov, Y. Hristova-Ivanova, A. Slavchev, Z. Denkova, G. Kostov (2019). Chemical composition, antioxidant activity and antimicrobial activity of essential oil from *Citrus aurantium* L zest against some pathogenic microorganisms. Zeitschrift für Naturforschung C. *A Journal of Biosciences, volume* 74, issue 5-6.

13) Kostov G., V. Shopska, R. Denkova-Kostova, B. Goranov (2019). Analytical approaches to determine the specific biomass growth rate in brewing. *Proceedings 33rd European Conference on Modelling and Simulation ©ECMS*, pp. 125-131.

14) Slavov A., P. Denev, Z. Denkova, G. Kostov, R. Denkova-Kostova, R. Chochkov, I. Deseva, D. Teneva (2019). Emerging cold pasteurization technologies to improve shelf life and ensure food quality. In: *Food Quality and Shelf Life* (Ed. Charis M. Galanakis), Academic press, Elsevier, p. 55 – 123. ISBN: 978-0-12-817190-5.

15) Yantcheva, N., I. N. Vasileva, P. N. Denev, H. N. Fidan, R. S. Denkova, A. M. Slavov (2019). Utilization of essential oil industry chamomile wastes as a source of polyphenols. *Bulgarian Chemical Communications,* Volume 51, Special Issue D (pp. 178 – 183).

16) Petelkov, I., V. Shopska, R. Denkova-Kostova, G. Kostov, V. Lyubenova, (2020). Investigation of Different Regimes of Beer Fermentation with Free and Immobilized Cells. *Periodica Polytechnica Chemical Engineering.* 64, 2 (2020), 162-171. https://doi.org/10.3311/PPch.13845.

Kristina Atanasova Ivanova

Affiliation: Institute of Canning and Food Quality, Agricultural Academy, Plovdiv, Bulgaria

Education:
2008 – 2012 – Bachelor; Technology of Tobacco and Tobacco products; University of Food Technologies
2013 – 2014 – Master; Food Safety; University of Food Technologies;
2015 – 2017 - PhD; 5.12 Food technologies.

Business Address: 154 Vasil Aprilov Blvd, Plovdiv, Bulgaria

Research and Professional Experience: Food technologies, Sustainability, Extrusion, Brewers' spent grains

Professional Appointments: Chief Assistant Professor

Honors: PhD

Publications from the Last 3 Years:
1) Petrova, T., Manev, Z., Bakalov, I., Ivanova, K., Ruskova, M., Penov, N., (2019). Structural-mechanical characteristics of fruit jam enriched with fibers, *AIP Conference Proceedings*, 2075 (1), https://doi.org/10.1063/1.5091361.
2) Petrova T. V., M. M. Ruskova, N. D. Penov, A. T. Simitchiev,, I. Y. Bakalov, V. B. Nenov, M. M. Momchilova (2019). *Optimization of extrusion process for production of apple pomace - wheat semolina extrudates. Health, Physical Culture and Sport,* 4(15), 212-219; ISSN: 2414-0244.
3) Kostov, G., R. Denkova-Kostova, V. Shopska, P. Nedyalkov, Z. Denkova, B. Goranov, V. Iliev, K. Ivanova, D. Teneva, New approach to modelling the kinetics of the fermentation process in cultivation of lactic acid bacteria, *Proceedings of 32nd European Conference on Modelling and Simulation ECMS* 2018, 212-218, ISSN: 2522-2414, https://doi.org/10.7148/2018-0212.
4) Iserliyska D., G. Zsivanovits, B. Gechev, M. Marudova, T. Petrova, M. Ruskova, I. Bakalov, A. Iliev (2019). Influence of Different Polysaccharides on Rheological Behavior and Thermal Properties

of Gluten Free Dough. *AIP Conference Proceedings*, 2075 (1), 160018. https://doi.org/10.1063/1.5091345.
5) Bakalov, I., T. Petrova, M. Ruskova, K. Ivanova, A. Simitchiev, St. Madjarova (2019). Influence of conditions of extraction of brewer`s spent grain with wheat semolina on the structural mechanical properties of the received extrudates. *Proceedings of 3rd International Conference Biosys Food Eng* 2019, E347, 6 pp; SBN 978-963-269-878-6. http://biosysfoodeng.hu/USB/pdf/E347.pdf.

Georgi Atanasov Kostov

Affiliation: Department of Technology of wine and brewing, University of Food Technologies, Plovdiv, Bulgaria

Education:
1999 – 2003 – Bachelor; Mechanical engineering and equipment; University of Food Technologies;
2005 – 2006 – Master; Mechanical engineering and equipment; University of Food Technologies;
2003 – 2007 - PhD; 02.01.24. Machines and apparatus for food and flavor industry; University of Food Technologies;
2007 – Assistant Professor;
2011 – Associated Professor;
2015 – Doctor of Science; 5.12. Food technologies;
2020 – Professor, 5.12. Food technologies;

Business Address: 26 Maritza Blvd, Plovdiv, Bulgaria

Research and Professional Experience: Bioreactors; Kinetics of fermentation process, Beer production with immobilized cells; Modelling of beer fermentation; Beer technology; Antioxidant activity of malt and wort; Functional wort-based non-alcoholic beverages; Functional food development and modeling.

Professional Appointments: Professor

Honors: DSc

Publications from the Last 3 Years:
1) Kostov G., I. Petelkov, R. Denkova, V. Shopska, Z. Denkova, B. Goranov, V. Iliev (2017). Modeling of Continuous Ethanol Fermentation In Ideal Mixing Column Bioreactor. *Proceedings 31st European Conference on Modelling and Simulation ©ECMS*, pp. 335-341.
2) Denkova, R., B. Goranov, Z. Denkova, D. Teneva, G. Kostov (2017). Enhancing yogurt health benefits: Development of starters for dairy and non-dairy yogurt. *Functional Foods: Sources, Health Effects and Future Perspectives.* Nova Science Publishers, Inc. Hauppauge, NY, 221 - 282.
3) Denkova, R., G. Kostov, Z. Denkova (2017). Functional bread: development of sourdough starters to improve bread quality. *Functional Foods: Sources, Health Effects and Future Perspectives.* Nova Science Publishers, Inc. Hauppauge, NY, 43 - 120.
4) Teneva D., R. Denkova, B. Goranov, Z. Denkova, G. Kostov (2017). Antimicrobial activity of Lactobacillus plantarum strains against Salmonella pathogens. *Ukrainian Food Journal. 2017. Volume 6. Issue 1: 125 – 133.*
5) Denkova R., B. Goranov, D. Teneva, Z. Denkova, G. Kostov (2017). Antimicrobial activity of probiotic microorganisms: mechanisms of interaction and methods of examination. In: *Antimicrobial Research: Novel bioknowledge and educational programs"* (*Microbiology Book Serie*s - Volume #6) 201 - 212.
6) Ivanova K., R. Denkova, G. Kostov, T. Petrova, I. Bakalov, M. Ruscova, N. Penov (2017). Extrusion of brewers' spent grains and application in the production of functional food. Characteristics of spent grains and optimization of extrusion. *J. Inst. Brew.*, 123 (4), pp. 544-552. doi 10.1002/jib.448.

7) Ahmedova G., M. Kaneva, R. Denkova, D. Teneva, Z. Denkova, I. Ignatov, P. Nedyalkov (2017). *Dynamics of the fermentation process of a lactic acid beverage based on wort and mint with* Lactobacillus casei *ssp.* rhamnosus *Oly.* Scientific Works of the University of Food Technologies 64(1).

8) Kostov G., R. Denkova-Kostova, V. Shopska, P. Nedyalkov, Z. Denkova, B. Goranov, V. Iliev, K. Ivanova, D. Teneva (2018). New approach to modelling the kinetics of the fermentation process in cultivation of lactic acid bacteria. *Proceedings 32nd European Conference on Modelling and Simulation ©ECMS*, pp. 212-218.

9) Vasileva I., R. Denkova, R. Chochkov, D. Teneva, Z. Denkova, Tz. Dessev, P. Denev, A. Slavov (2018). Effect of lavender *(Lavandula angustifolia)* and melissa (*Melissa Officinalis*) waste on quality and shelf life of bread. *Food Chemistry, 253*(1): 13-21.

10) Denkova-Kostova R., B. Goranov, D. Teneva, Z. Denkova, G. Kostov (2018). Antimicrobial activity of *Lactobacillus* strains against *Escherichia coli*: a multimethod approach to explore the mechanisms and factors determining the antimicrobial action. In: *Antimicrobial Research: Novel bioknowledge and educational programs"* (Microbiology Book Series - Volume #7) (Enrique Torres-Hergueta and A. Méndez-Vilas Eds.), Microbiology Book Series, 7, 43 – 54, ISBN: 978-84-947512-5-7.

11) Lyubenova V., M. Ignatova, G. Kostov, V. Shopska, E. Petre, M. Roman (2018) An Interactive Teaching System for Kinetics Modelling of Biotechnological Processes, *Proceedings of 22nd International Conference on System Theory, Control and Computing (ICSTCC)*, 10-12 October 2018, Sinaia, Romania, 366-371 doi: 10.1109/ICSTCC.2018.8540688.

12) Nedyalkov P., R. Denkova, D. Teneva, V. Shopska, B. Goranov, Z. Denkova, G. Kostov, M. Kaneva (2019) Yeast selection for non-alcoholic and low-alcoholic beverages based on wort. *Food Science and Applied Biotechnology,* 2019, 2(2), 140-148 https://doi.org/10.30721/fsab2019.v2.i2.

13) Teneva D., R. Denkova-Kostova, B. Goranov, Y. Hristova-Ivanova, A. Slavchev, Z. Denkova, G. Kostov (2019). Chemical composition, antioxidant activity and antimicrobial activity of essential oil from *Citrus aurantium* L zest against some pathogenic microorganisms. Zeitschrift für Naturforschung C. *A Journal of Biosciences,* volume 74, issue 5-6.
14) Kostov G., V. Shopska, R. Denkova-Kostova, B. Goranov (2019). Analytical approaches to determine the specific biomass growth rate in brewing. *Proceedings 33rd European Conference on Modelling and Simulation ©ECMS, pp. 125-131.*
15) Slavov A., P. Denev, Z. Denkova, G. Kostov, R. Denkova-Kostova, R. Chochkov, I. Deseva, D. Teneva (2019). Emerging cold pasteurization technologies to improve shelf life and ensure food quality. In: *Food Quality and Shelf Life* (Ed. Charis M. Galanakis), Academic press, Elsevier, p. 55 – 123. ISBN: 978-0-12-817190-5.
16) Kostov G., V. Shopska, R. Denkova-Kostova, B. Goranov (2019). Analytical approaches to determine the specific biomass growth rate in brewing. *Proceedings 33rd European Conference on Modelling and Simulation ©ECMS, pp. 125-131.*
17) Shopska V., R. Denkova, V. Lyubenova, G. Kostov (2019) Kinetic Characteristics of Alcohol Fermentation in Brewing: State of Art and Control of the Fermentation Process. In: *Fermented Beverages* Volume 5: The Science of Beverages (Edited by A. M. Grumezescu, A. M. Holban), Woodhead Publishing, 529-576 https://doi.org/10.1016/B978-0-12-815271-3.00001-4 ISBN: 978-0-12-815271-3 (print) ISBN: 978-0-12-815703-9 (online).
18) Shopska V, M. Dzivoderova-Zarcheva, K. Ivanova, P. Nedyalkov, R. Denkova-Kostova, M. Kaneva, G. Kostov (2019) Phytochemical characteristics of different malts and possibilities for their application in functional beverages. *Proceedings of the X International Scientific Agricultural Symposium "Agrosym 2019,"* 03-06 October 2019, Jahorina, Bosnia and Herzegovina, 785-790 ISBN 978-99976-787-2-0.

19) Kostov, G., V. Iliev, B. Goranov, R. Denkova, V. Shopska (2020) Immobilized cell bioreactors in fermented beverage production: design and modelling, In: 19. *Biotechnological progress and beverage consumption, The Science of Beverages series* (A. M. Grumezescu and A. M. Holban eds.), 339-376. https://doi.org/10.1016/B978-0-12-816678-9.00011-4.

20) Petelkov, I., V. Shopska, R. Denkova-Kostova, G. Kostov, V. Lyubenova, (2020). *Investigation of Different Regimes of Beer Fermentation with Free and Immobilized Cells. Periodica Polytechnica Chemical Engineering.* 64, 2(2020), 162-171. https://doi.org/10.3311/PPch.13845.

In: Beer: From Production to Distribution
Editor: Armand Legault

ISBN: 978-1-53618-414-3
© 2020 Nova Science Publishers, Inc.

Chapter 2

THE REVOLUTION OF CRAFT BEER

Cristina Calvo-Porral[1], and Sergio Rivaroli[2]*
[1]Department of Business Administration,
University of A Coruña, A Coruña, Spain
[2]Department of Agricultural and Food Sciences,
University of Bologna, Bologna, Italy

ABSTRACT

Craft beer has received much attention over the past two decades. The stability of consumption of beers and the rising interest towards craft beers is a clear signal that consumption patterns are changing. As a consequence, craft breweries have emerged as an alternative to industrial beer in many countries. The limited production and the use of unique non-traditional ingredients during the production process offer beers with original flavors, aromas, textures, and styles that are welcomed by consumers. Moreover, drinking craft beer is perceived as a sensory consumption experience and way to move away from industrial beer and to explore new taste experiences. So, the phenomenon of craft beer should be considered and analyzed. In this context, this chapter aims to describe the so-called "craft beer revolution," shedding some light on some of the most significant

---

* Corresponding Author's Email: ccalvo@udc.es.

drivers of craft beer consumption that are changing the status quo of the alcoholic beverage sector. More precisely, this study aims to provide a conceptualization and characterization of craft beer and a full description of the "craft beer revolution" phenomenon, explaining the increasing consumption of this beverage. Besides, the chapter will analyze the craft beer production and distribution systems, characterized by its limited production, the use of non-traditional and unique ingredients, and a specific trade-off consumption. Finally, the chapter will examine the main drivers and motivations underlying craft beer consumption.

Keywords: craft beer, revolution, quality, flavour

INTRODUCTION

According to the World Health Organization (WHO, 2018), beer is one of the most consumed alcoholic beverages worldwide, being part of the daily consumption of millions of people. Among beer consumption, in the last years, the beer industry has experienced a significant increase in the craft beer demand and consumption, which could be described as the "craft beer renaissance".

What are the reasons for this phenomenon?. Today, consumer beer preferences appear to be associated with the search for higher product quality and product craftsmanship, as well as growing demand for new exciting beer flavors and styles. As a consequence, these factors are gradually modifying beer consumption habits, preferences, and values. The modification in beer consumption habits represents a clear signal of an increasingly profound and conscious beer culture spreading worldwide, which could be interpreted as a "revolution". Further, this increasing movement of beer consumers who are most inclined to consume and taste craft beer allows the denomination of this revolution as a "craft beer revolution".

So, today beer consumers are demanding new beer styles and joining the "craft beer revolution". These craft beer consumers have become more aware of the emerging beer culture, in terms of quality and consumption habits, being more aware of the nutritional components of beer and the health benefits derived from beer consumption (Aquilani et al., 2015). Similarly, these new beer consumers prefer to drink high-quality beer products, instead of industrial beers, despite the higher prices.

Therefore, craft beer is a niche product intended for beer involved consumers. Consequently, this "craft beer revolution" has created a new market opportunity for craft beers, which have different sensory attributes and characteristics compared to industrial beers that facilitate insightful beer product differentiation.

So, considering that craft beer is becoming more and more popular among beer consumers and that the craft beer revolution seems to be extending worldwide, the present chapter aims to analyze all the relevant aspects of this phenomenon.

BEER AS AN ALCOHOLIC BEVERAGE

The beer industry has a long history that begins with the first beer made by Sumerians 5,000 years ago (Kiefer, 2001). Later, the European monks contributed to beer product diversification introducing some product innovations. Before the XIX century, beers were made in small batches by individual small brewers, and there was no control over ingredients or production processes. As a consequence, each beer was unique and highly differentiated from all other beers and unlikely to be reproduced in a similar form or flavor (Cardello et al., 2016).

However, it was during the Industrial Revolution that the beer available in the market achieved a consistent level of quality (Jaeger et al. 2017a), and it was possible to mass-produce beers of standard quality. This industrialized beer production led to the consolidation of breweries.

Consumers perceive beer as a soft alcoholic beverage that helps as a thirst quencher, for social interaction, or even as an ingredient of more complex preparation (Silva et al., 2016; Gómez-Corona et al., 2017c). Further, beer as a thirst quencher is strongly associated with informal and relaxing occasions, being a symbol of demarcation between work and non-work hours (Pettigrew and Charters, 2006). Additionally, beer sensory attributes, such as perceived quality, carbonation, aroma, or taste, as well as the context and moment of consumption, are factors that influence beer preferences incorporated into the consumer's mental representation of beer as an alcoholic beverage (Aquilani et al., 2015).

The beer category is strongly dominated by one style and to a broader range of options of industrial beers and less commercial opportunities such as craft beers (Gómez-Corona et al., 2017a). What is more, large brewing companies reinforced their dominant position in the marketplace with the proliferation of mass-produced beer products. The consumption of mass-produced industrial beer could be considered as a low-intensity consumption experience, since industrial beer is a popular product, with no outstanding special characteristics, that can be seen as a commodity of the beverage category (Gómez-Corona et al., 2016).

However, in the last decades, beer consumers began to avoid mass-produced industrial beer. As a consequence, the beer market experienced a contraction in demand for industrial beer production in favor of premium specialty beers (Jaeger et al., 2017a). In this new context, micro-breweries began to emerge (Jaeger et al., 2017a).

Beer Consumption Behavior

Beer is one of the most widely consumed alcoholic beverages worldwide, accounting for about 78.2% of the alcohol beverage drink shares (Gómez-Corona et al., 2016), and it is always present on special occasions or in specific contexts (Gómez-Corona et al., 2017c). Besides, some authors indicate that beer consumption is also becoming more and more fashionable, trendy, and connected to modern lifestyles (Gómez-Corona et al., 2017a).

Following Gutjar et al. (2015), the individuals' response to a food or beverage product does not only depend on the product itself, namely its intrinsic and extrinsic attributes, but also on the product associations in terms of functional and emotional associations.

On one side, the beer functional conceptualization could be its perception as being "a thirst quencher" and as being a symbol of demarcation between work and non-work hours, in eating and non-eating social contexts (Pettigrew and Charters, 2006). Likewise, today consumers pay great attention to the nutritional components and health benefits associated with moderate beer consumption (Aquilani et al., 2015). However, the emotional attributes and perceptions of beer are the most important associations (Silva et al., 2016).

On the other side and regarding the emotional conceptualization of beer, prior research highlights that "feeling relaxed" is probably the most prevalent association with beer (Charters and Pettigrew, 2008). Similarly, beer and other alcoholic beverages are often consumed in situations that have significant emotional components, such as consumption at parties, at sporting events, or during special evening meals (Cardello et al., 2016). Regarding consumers' feelings before beer consumption, previous research reports that beer is consumed either to enhance the intensity of positive emotions and to decrease the negative ones (Silva et al., 2016). More precisely, this alcoholic consumption pattern is known as "emotionally instrumental drinking", which means drinking with the purpose to control both positive and negative emotions, being a significant motivation for alcohol consumption. Finally, and considering craft beer, some authors indicate that substantial emotional aspects are related to craft beer consumption, given that this product is perceived as made with dedication and great affection (Koch and Sauerbronn, 2019).

How is the sensory experience of beer consumption?. According to Silva et al. (2016), the eating and drinking experience is influenced by two processes in the consumers' minds: the identification of the product through sensory perception, and the establishment of associations assigned when thinking about a specific food or beverage. In this context, the beer sensory experience can be understood as the effect the beer has on a consumer, derived from direct sensory or multisensory stimulation (Gómez-Corona et al., 2017c). According to Silva et al. (2016), beer consumption behavior is mostly driven by consumers' unconscious minds, being strongly associated with a positive emotional impact related to the experience. Following Aquilani et al. (2015), the most important sensory attributes of beer are its perceived quality, aroma, and carbonation, as well as the context and moment of consumption, culture, character, and affect.

There are differences between the industrial and craft beer sensory consumption experience?. Prior studies point out that the industrial beer consumption experience should be refreshing and should have a "thirst soothing" effect. Therefore, this consumption experience seems to be focused on the functional attribute of beer as a "thirst quencher", rather than a sensory consumption experience based in enjoyment (Gómez-Corona et al., 2016). Conversely, craft beer consumer experience is described in the literature as a combination of a sensory and cognitive experience that offers pleasure, enjoyment, social recognition, and a sense of identity (Gómez-Corona et al., 2016). This consumption experience has also been described as a symbolic product consumption experience (Gómez-Corona et al., 2016). One possible explanation for these differences between industrial and craft beer may be that craft beer consumers take the time to feel and enjoy the sensory characteristics of beer and to think about the consumption experience itself (Gómez-Corona et al., 2017c), but industrial beer consumers do not.

Finally, regarding the industrial and craft beer consumption, it should be noted that even though beer consumption is mostly dominated by industrial beer, craft beer is gaining market share worldwide with increasing popularity among consumers (Aquilani et al., 2015). Nowadays, craft beer could be considered as a product category itself, increasingly moving away from the industrial beer, which has been the only beer available in the marketplace for decades (Gómez-Corona et al., 2017c). Thus, craft beer is already changing the way consumers perceive the beer product category, being one significant consequence that craft beer drinkers have a completely different consumption pattern compared to industrial beer drinkers.

CRAFT BEER: A DESCRIPTION OF THE PHENOMENON

As reported by the World Health Organization (WHO, 2018), 34.3% of all recorded alcohol consumed in the world is beer, representing the second most consumed alcoholic beverage after spirits.. Based on the same source of information, in 2016, beer still represented the most popular alcoholic beverage in America (53.8%) and Europe (40.0%) where the share of the total consumption of beer increased by two percentage points compared to 2015.

Among beer, craft beers have revolutionized the market of this alcoholic beverage since the end of the twentieth century, contributing to the take-off of the beer sector after a long period of consolidation. Interestingly, the origin of craft beer could be traced back to the 70s in the United States (Fastigi et al. 2015). Then, and according to Garavaglia and Swinnen (2018b), the increasing income of people after World War II, the changing attitudes of consumers (i.e., the so-called "neo-localism" and the demand for variety), and the growing availability of technical equipment and capital for small breweries have been favorable conditions for the resurgence of craft breweries from the 1980s onwards. Other authors, such as Donadini and Porretta (2017), highlight that industrial beer drinkers have mainly fueled the steady gain in the market share of craft beer. As a consequence, the astonishing rise of the number of craft breweries over the past decades contributed to revitalizing the beer market, and still attracting the attention of scholars from different disciplines to understand what will be the future of the sector.

The Concept of Craft Brewery

Even though craft beer has received much consideration in recent years, to now, there is still not a broadly accepted definition or classification of the term "craft brewery" or "craft beer". Sometimes terms such as "microbrewery" or "artisanal brewery" are used to differentiate local, traditional, small, and independent breweries, from the few global multinationals that dominate the space of the beer market. However, some consumers do not understand what is meant by craft beer and also find it difficult to tell which brands are craft beers (Gómez-Corona et al., 2016).

In this context, some authors of the marketing area have made some attempts to conceptualize the term "craft beer". For example, according to Acitelli and Magee (2017), craft beer can be defined as "beer produced by any small, independently owned brewery that adheres to traditional brewing practices and ingredients". This definition comprises the two most essential elements of craft beer, namely, the traditional nature of the beer production method, and the small size of the production facility in which it is brewed. Therefore, craft beer stands in high contrast to commercial, industrial beers produced by large brewers in large batches, using standard ingredients and often containing additives (Elzinga et al., 2015).

Similarly, authors like Giacalone (2013) defined craft beer as a heterogeneous group of beers varying hugely in flavors for being brewed from a wide range of untraditional grains, fruits, herbs and spices, for which consumers' liking is maximized when the beer contains an optimum amount of novelty and does not deviate too much from consumers' expectations. Likewise, authors such as Garavaglia and Swinnen (2018a) propose some distinguishing features to define craft breweries: production process, ownership, production scale, age, and tradition. Thus, providing an overview of craft beer definition is a relevant premise to understand the different historical local cultures in beer brewing, and to have an appropriate interpretation of the craft beer sector worldwide.

Also, different countries have various legal or tax-related definitions of craft beer, and in this chapter, the definitions provided by the US, Italy, Spain and the Netherlands will be examined.

According to US standards, a "craft brewer" is a small, independent, and traditional brewer (American Brewers Association, 2016) that must satisfy three requirements. In the first place, the brewer should have an annual production of 6 million barrels of beer or less. In the second place, 25% of the craft brew be owned or controlled by an alcoholic beverage company that is not a craft brewer. Finally, the craft brewer should produce the majority of its total beverage volume of beers from traditional or innovative brewing ingredients or their fermentation.

Thus, the American Brewers Association (2020) defines a craft brewery as based on small production, independent brewing companies, and the use of traditional ingredients. Moreover, for this association, "small production" refers to a maximum annual production of 7,038,041 hectolitres (6,000,000 barrels) (American Brewers Association, 2020). Likewise, the American Brewers Association divides breweries into microbreweries and brewpubs based on the annual production. The former is defined as a brewery that produces up to 17,595 hectolitres (15,000 barrels), whereas a brewpub is considered a brewery that sells 25% or more of its beer on-site.

Similarly, in Italy, the Italian Parliament defines a craft brewery as "a brewery that is legally, economically and physically independent, with an annual production that does not exceed 200,000 hectolitres" (170,502 barrels) (Law No. 154/2016). Moreover, the Italian Senate defines the concept of craft beer describing it as "the product of a small scale and independent brewery, not subjected during production to micro-filtration and pasteurization". Furthermore, the Italian craft breweries association Unionbirrai -the Italian association that protects craft beer-, distinguish microbreweries, brewpubs, and beerfirms. The first is considered a brewery exclusively dedicated to the craft brewing, whose production is directed to restaurants, pubs, or shops. Brewpubs are considered companies that add consumption on the premises to production. Whereas beerfirms, also called "brewfirms", are companies where the production of beer is carried out in plants that are not the company's property. Recently a new form of production unit was introduced in Italy named "agricultural brewery", that is a brewery characterized by the obligation to produces at least 51% of the raw materials used during the brewing process.

In Spain, the Spanish Brewing Association defined craft breweries as beer companies holding their plants, which are not participated directly or indirectly by other companies in the sector, with a maximum production of 50,000 hectolitres (42,626 barrels), which not use ingredients other than barley and wheat malt as a source of starch, except for those beers that are characterized by the use of different raw materials.

Similarly, in the Netherlands, a craft brewery is defined as an independent brewery whose annual production does not exceed 1,000,000 hectolitres (852,510 barrels) (CRAFT, 2019). Bentzen and Smith (2018), instead, referring to Danmark, define the limit of 50,000 hectolitres (42,626 barrels) of beer produced annually as a descriptor to identify a craft brewery.

Thus, "independence", "tradition", and "volume" are essential features considered by national brewer organizations to release seals that allow brewers to highlight their credentials as craft beer producers. As illustrated above, the volume is one of the relevant descriptors to identify a craft brewery[1] and is strictly related to the size of the country. The different cut-off values make it challenging to conduct a comparative analysis of the craft brewing sector amongst countries worldwide. For example, as pointed out by Garavaglia and Swinnen (2018a), considering the US's cut-off criteria in Europe, most of the mass brewers would be considered craft brewers[2].

However, what is the consumer viewpoint?. Some authors point out that for the ordinary beer consumer craft beer is a well-differentiated product, that distinguishes itself from commercial industrially brewed beer (Donadini and Porretta, 2017). One of the reasons for its easy distinction is the unique, rich, and composite sensory profile of craft beer, which tends to be related to beer flavor. Another potential reason is the list of ingredients –some with a local origin- that may extend to include unconventional ingredients used for brewing. Finally, another explanation could be the beer production through small-scale breweries not subjected to pasteurization and filtration (Donadini and Porretta, 2017).

---

[1] In addition to the volume of beer, the terms "independent" and "traditional" are two additional distinctive aspects to define a craft brewery. The former refers to the ownership structure, whereas "traditional" refers to the nature of the beer production method. The ABA defines independent a brewery in which less than 25% is owned or controlled by an alcohol industry member that is not itself a craft brewer. Whereas "traditional" refers to the fact that 50% of beer derives the flavor from "traditional" brewing ingredients. Differently, in Italy, the concept of tradition is related to the production of a beer that does not undergo pasteurization and micro-filtration. What is worthy to note, is that the Italian definition of craft beer neither establishes limits on the use of adjuncts nor mentions the use of local or traditional ingredients to be used in the craft brewing process.

[2] A legal definition of craft beer is a question which should be addressed by all governments, especially in those countries who have a solid beer tradition. Considering the differences in terms of beer's culture worldwide, the best option could be the identification of national definitions, possibly based on shared and harmonized guidelines.

The Craft Beer Sector: Global Trends

Consumption of craft beer has been changing worldwide, moving from a specialty rarely known product to a typical product amongst consumers (Gómez-Corona et al., 2016).

Recent estimates put the total percentage of craft beer at 6% of the global beer market (Ascher, 2012). According to Donadini and Porretta (2017), the fast growth experienced by craft beer has benefited from the structure of the actual beer industry. Schnell and Reese (2003), indicate that the main competitive advantage of craft beer depends on the production of ancient beer styles no longer produced by large brewing companies, as well as on the association between the product and the territory. Similarly, Donadini et al. (2016) report that the main advantage of craft beer in the marketplace is based on its creativity, authenticity, and the experimentation of innovative combinations of ingredients and flavors. As pointed out by Donadini et al. (2016), nowadays large industrial breweries are taking notice of this new trend and are trying to reposition themselves with an increased focus and investment on craft beer as part of a diversification strategy.

Although the craft beer sector is receiving much attention, official statistics regarding the craft beer market are not always available, and generally are limited to the number of brewing companies. Nonetheless, the evolution of the number of microbreweries and brewpubs can provide us with a clear picture of the rising vitality of the craft beer market. Thus, to simplify matters, the term craft brewery is used in the following paragraphs to describe microbreweries as well as brewpubs that produce beer. In this light, one source of information is related to the website www.RateBeer.com (referred to here as "RateBeer"). RateBeer, is considered one of the most-visited sources for beer and craft beer information, which also registers and updates the number of active microbreweries and brewpubs worldwide.

Figure 1 illustrates the rapid growth of the global number of craft breweries, starting with 361 companies in 1990 and reaching more than 20.7 thousand craft breweries in 2019, with an average annual growth rate (referred to here as AGR) of 15%. Europe and America dominate the craft beer sector over the period, representing 49.1% and 39.8% of the craft breweries worldwide in 2019, respectively. It is worthy to note the rapid growth of the Asian craft beer sector; from 1990 to 2019, the number of craft breweries increased with an AGR of 14.4%, resulting in 1,461 companies in 2019.

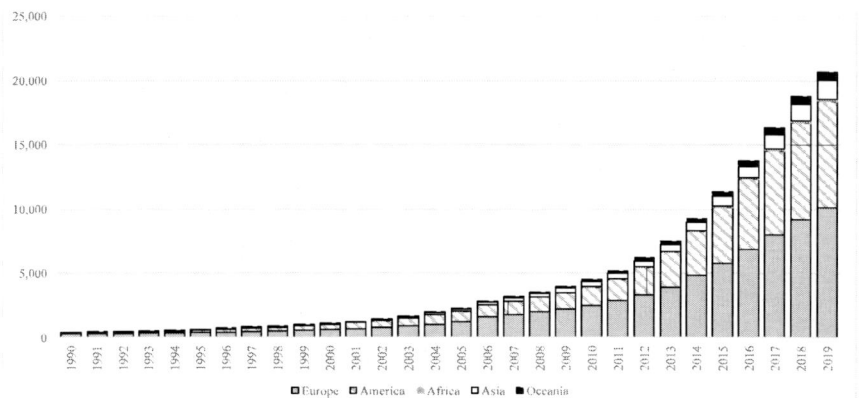

Source. Elaboration on RateBeer.com.
Note. The number of craft breweries includes microbreweries and brewpubs.

Figure 1. Number of craft breweries worldwide by continents (1990-2019).

The same evolution occurred in Oceania and Africa, with the number of new craft breweries always on the rise, even though overall, they represented only 3.1% of the global craft breweries in 2019. An overview of the composition of the worldwide craft beer sector highlights that the quota of microbreweries declined during the 1990-2006 period. Only after 2006, this situation changed with a resurgence of the quota of microbreweries (Figure 2).

An analysis of the evolution of the number of craft breweries in America brings to light that the United States dominates the scenario, representing 71.6% of all the craft breweries censused by RateBeer in 2019. The growth of the US's craft breweries has been spectacular, with an AGR of 17.9% from 1990 to 2019 (Figure 3). According to Elzinga, Tremblay, and Tremblay (2015), this growth finds it rooted in the 1960s when Fritz Maytag purchased the Anchor Beer Company (U.S.A.). What is more, it results in the so-called "craft beer revolution" that spread across Europe and, slightly, also in China and Australia (Danson et al. 2015; Cabras and Bamforth 2016; Elzinga, Tremblay, and Tremblay 2015; Argent 2018). Among the remaining states of North America, Mexico represents 72.2% of the craft breweries in 2019 ($AGR_{2000\text{-}2019}$=12.9%). Brazil, Argentina, and Chile host 68% of South American craft breweries in the last year investigated.

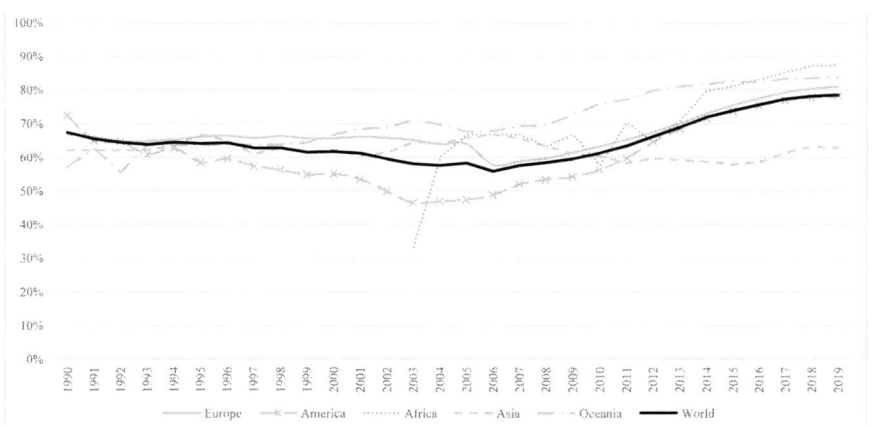

Source. Elaboration on RateBeer.com.

Figure 2. The quota of microbreweries worldwide, 1990-2019 (%).

A European cross-country analysis reveals that in 2019 30.4% of European craft breweries were concentrated in four countries with a long tradition in brewing, such as England (13.9%; $AGR_{1990\text{-}2019}$=12.9%), Germany (9.4%; $AGR_{1990\text{-}2019}$=7.6%), Czech Republic (3.6%; $AGR_{1990\text{-}2019}$=16.9%) and Belgium (3.5%; $AGR_{1990\text{-}2019}$=10.1%). Although this finding reveals that the craft beer revolution is deeply rooted in countries with a tradition in the brewing industry (Figure 3), it is noteworthy that other European countries without a long-brewing background have been riding this phenomenon as well. For example, in France, craft breweries have grown by 15.1% from 2010 to 2019, reaching almost 1,200 companies in 2019. It is worth pointing out that in the 28 European member states (EU28), France is the third-largest producer of malt after Germany and the UK (Euromalt, 2020). According to previous literature on food marketing, it can be stated that craft beer has gained a top market share in Europe due to a minority of highly involved beer consumers. What is more, for these consumers, the relatively limited distribution and availability of craft beer and the influence of higher prices were not a big concern (Giacalone, 2013).

A cross-country analysis of the Asian craft beer sector reveals that 68.9% of companies are located in Japan (46.3%), Thailand (9.%), India (7.0%) and China (6.6%), whereas in Oceania, 97.7% of craft breweries are located in Australia (73.2%) and New Zeland (24.5%).

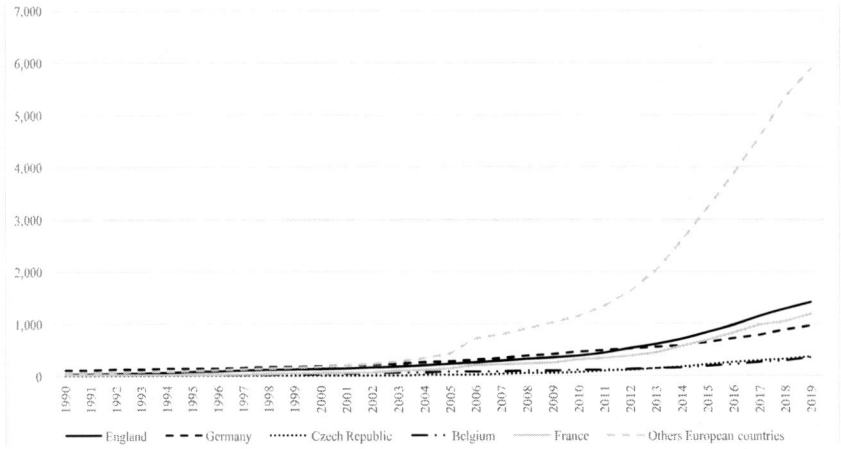

Source. Elaboration on RateBeer.com.
Note. The number of craft breweries includes microbreweries and brewpubs.

Figure 3. Number of craft breweries in four European countries with a long tradition in the brewery (1990-2019).

The Craft Beer Sector in Europe

According to Berkhout et al. (2013), Europe was the second-largest beer producer worldwide in the year 2012, with a production of 390 million hectoliters of beer, and an amount of 357 million hectoliters consumed. Further, Europe has experienced an increase in the number of breweries reaching the number of 5,000 with total annual sales of approximately 111 billion euros.

In general terms, it can be stated that Europe has been somewhat slower to develop a craft beer industry, compared to the US (Jaeger et al., 2020). However, in Europe, the craft beer sector is one of the fast-growing segments in the beverage industry (AssoBirra, 2014), with beer sales growth being primarily due to the success of premium and craft beers at the expense of standard products (AssoBirra, 2014). Besides, and regarding craft beer distribution, Europe has experienced an increase in off-trade consumption through modern and traditional distribution channels, instead of an increase on on-trade consumption in the hospitality sector such as in restaurants, breweries or pubs (AssoBirra, 2014).

What is the reason for the increase in the craft beer sector in Europe?. Some authors support that craft beer has been experiencing an astonishing boom in the last two decades and continue expansion in Europe due to consumers' increasing demand (Donadini et al., 2016). What is more, according to Aquilani et al. (2015), beer consumers are particularly interested in tasting new craft beers with different flavors, aromas, and styles that move away from mass-produced industrial beers. Finally, some authors notice that nowadays, European beer consumers drink less beer. Still, they indulge themselves more with special and pricier beers since they are getting more affluent and adventurous, as well as more conscious about beer quality (Donadini et al., 2016).

The Craft Beer Sector in Italy

From the mid-90s, craft beer started to spread in the Italian market, mostly in the Northern regions of the country. Subsequently, the popularity of craft beer and the number of microbreweries boomed across Italy, moving from around 40 brewers in 2005 to more than 700 today (AssoBirra, 2014; Donadini and Porretta, 2017). The diffusion of craft beer in the last years has resulted in increased product diversity in the Italian beer market in terms of beer styles, flavors, and textures (Donadini and Porretta, 2017). Nowadays, the Italian craft brewing sector is still a relatively small economic niche. According to Cannatelli et al. (2015), it is dominated by small breweries with an average production that varies widely in the volume and variety of beer produced, in the type of the brewing company, in the location of the brewing plant and in the geographical territory where the brewer operates.

Besides, Italians are small beer consumers with an annual per capita consumption of beer far below the European average. Still, they are keen on craft beers, since craft beer is perceived as a high standard quality beer (Aquilani et al., 2015).

According to The Brewers of Europe's (2019), which is the principal representative institution of the European brewing sector, Italy ranks 4th in the hierarchy of the EU28 member states by numbers of craft breweries (Table 1). The expansion of the craft beer sector in a traditional "wine-producing" and "wine-consuming" nation, is as peculiar as spectacular.

Table 1. Craft breweries in the EU28 (2014-2018)

	2014	2015	2016	2017	2018	AGR (%) 2014-2018	2018 (%)	Ranking 2018
United Kingdom	1,378	1,527	1,817	1,878	1,978	9.5	27.6	1
France	566	690	850	1,000	1,450	26.5	20.2	2
Germany	682	723	738	824	853	5.8	11.9	3
Italy	505	540	718	693	692	8.2	9.6	4
Others EU28	*1,144*	*1,342*	*1,728*	*2,020*	*2,201*	*17.8*	*30.7*	
EU28	4,275	4,822	5,851	6,415	7,174	13.8	100.0	

Source. Elaboration on TheBrewersOfEurope.org.

One source of information for the Italian craft beer sector is related to the website www.microbirrifici.org (referred to here as Microbirrifici.org), which can be considered as one of the most accurate online databases with regards to the national microbreweries, brewpubs, and beerfirms. Based on this source of information, in Italy, the so-called "craft beer revolution" is a more recent phenomenon that has been rapidly growing since the mid-2000s (Figure 4). According to the Microbirrifici.org, from 1996[3] to 2019, the craft breweries involved in this sector (microbreweries and brewpubs) increased with an AGR of 24%, resulting in 982 companies in 2019. Besides, it is

---

[3] The year 1996 was considered as initial year of the Italian craft brewing sector, because in 1996 was approved by Italian Parliament the definition of artisanal beer.

worth pointing out that the number of beerfirms went from 1 to 585 between 2005 and 2019 ($AGR_{2005-2019}=26.3\%$).

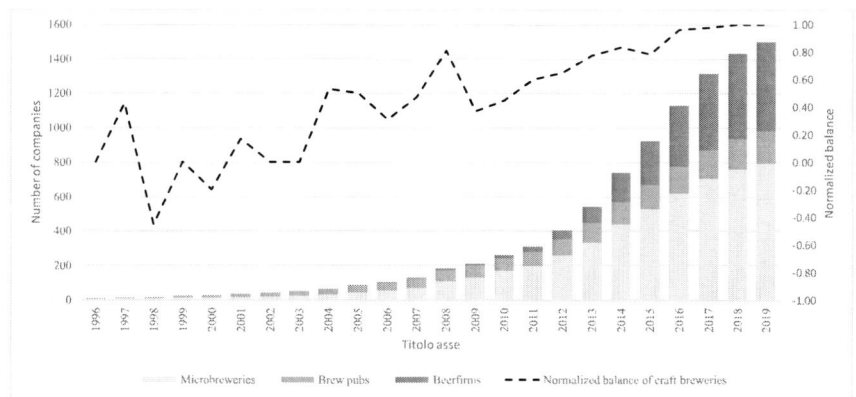

Source. Elaboration on Microbirrifici.org.

Note. The normalized balance is calculated as craft breweries' entries minus exits, divided by the total amount of entrances and exits.

Figure 4. Active microbreweries, beerfirms, and brewpubs in Italy (1990-2019).

As pointed out by Fastigi, Viganò, and Esposti (2018), this impressive evolution raises questions about when the craft beer euphoria phase will give way to a stage of consolidation of the Italian craft beer sector. A look at the distribution of entries and exits of craft breweries over 1996-2019 highlights a slowdown of the so-called "craft beer euphoria" after 2014. Although the numbers of craft breweries were growing steadily over the period, there has been a reduction of entries and exits during the last five years. Thus, this trend could suggest that the Italian craft beer sector has been living a phase of a gradual consolidation since 2015.

Based on the updated information provided by AssoBirra (2014), the Italian Association of Malters and Brewers, the Italian beer consumption from 2008 to 2012 remained stable at approximately 30 liters per capita, being a low value compared to other European countries. Later, in the period 2014-2018, the market share of craft beer grew from 2.4% to 3.2%. During the same period, the production of Italian craft beer has increased from 378 to 504 thousand hectolitres, resulting in an average yearly production of 545 hectolitres per craft brewery. Furthermore, during the same period, each craft brewery employed, on average, four employees every year (AssoBirra, 2020).

Interestingly, and according to the data provided by AssoBirra (2014), the Italian consumer behavior changed, paying more attention to quality. More precisely, super-premium beers and specialty beers have increased their market share in the Italian market, due to the increased consumers' curiosity regarding craft products and new beer styles (Berkhout et al., 2013). Further, these authors suggest that Italian consumers have become more inclined to drink less quantity of beer, while spending the same amount of money for more expensive quality products, such as premium or specialty beers (Berkhout et al., 2013).

Regarding the way of consumption, according to Menghini et al. (2019), 51% of craft beer in 2018 was sold in barrels, 41% in bottles, whereas the remaining 8% in a glass in the brewpubs. During the same year, two-thirds of craft beer was distributed within regional borders, 30% in other Italian regions, and only 3% of the product was sold abroad. When considering distribution channels, the direct channel is the most widespread. Cannatelli and Pedrini (2012) reported that 51.1% of craft beer sales come from direct distribution, 15.9% from the indirect distribution, and large scale retailing, and almost 31% of craft beer sales come from serving by the glass. Furthermore, Ravelli and Pedrini (2015) revealed as e-commerce is a distribution channel that contributes to a limited extent to the craft breweries' sales. What is more, craft breweries consider the large scale retailing as an opportunity to saturate their production capacity.

The Craft Beer Sector in Spain

Spain is an excellent example of the change in the beer consumption patterns, being characterized by the most remarkable change in the market share of total alcohol consumption in the last decades. During the decade of the 60s, the alcoholic beverage with the highest market share in consumption was wine, accounting for 65.4%, followed by spirits with 23.6% and beer with 11%. However, in the year 2005, beer was the alcoholic beverage with the highest market share, accounting for 48%, followed by a wine with 38.1% and spirits with 13.9% (Colen and Swinnen, 2011). Thus, it can be stated that in the decade of the 90s began a gradual change in the attitude of consumers towards the consumption of beer. Today, Spanish beer consumers are above average beer consumers in Europe, being particularly interested in off-trade beer consumption.

Despite the large volume of beer production in Spain, traditionally, beer variety has always been scarce (Garavaglia and Castro, 2018). It seems that beer production and consumption in Spain have adequated on a situation where, on the one hand, brewers offered a "thirst-quenching", cold, uncomplicated drink. On the other hand, consumers were enough satisfied with ordering and consuming an anonymous "caña" –a glass of draft beer-, without even asking what beer is being served (Garavaglia and Castro, 2018).

However, this situation has changed in the decade of the 90s, with few attempts of craft beer production, that has developed a more significant phenomenon in recent years. Following Gracia-Arnaiz (2011), in the late 90s, beer became to be considered as a "social drink" to be consumed for evenings and started to be accompanied to meals. The old type of beer consumer, identified as men of the popular classes, had changed; and in the decade of the 90s, the new beer consumer had a higher social status.

The first craft brewery located in Spain was Naturbier, a brewpub that opened in 1986 in Madrid. The main difficulty faced by this company was that consumers did not know about craft beer and did not understand the turbidity caused by yeast (Garavaglia and Castro, 2018). However, this was a successful experience, becoming an example for other brewing companies, such as the Barcelona Brewing Company. Interestingly, the microbrewers proved to Spanish consumers that there were many other existing varieties of beer, different from the "caña".

Also, the diffusion and communication of consumers' associations helped in spreading a new beer culture and attitude towards beer among young consumers; and this factor helped in fostering the demand for a greater variety of beers among consumers (Garavaglia and Castro, 2018). Accordingly, the number of craft breweries increase substantially (Figure 5), as well as the annual production volume of craft beer in Spain.

Similarly, new craft beer consumers emerged, who were willing to experiment with new beer products, and who had different attitudes and product preferences. This niche of consumers also wanted to participate in product creation directly (Garavaglia and Castro, 2018).

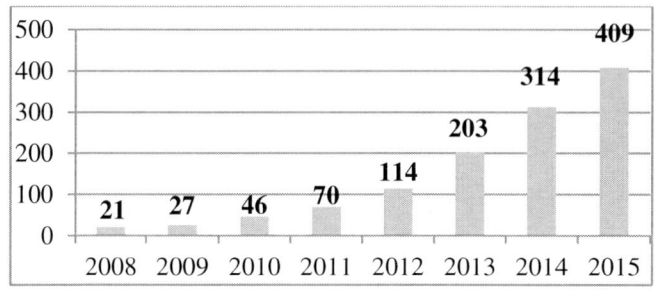

Source: Garavaglia and Castro (2018).

Figure 5. Increase of craft breweries in the Spanish market (2008-2015).

A PROFILE OF THE CRAFT BEER CONSUMER

How Is the Craft Beer Consumer?

As a consequence of the growth of craft beer and the significant shift in consumer beer preferences, a new segment of beer drinkers has emerged, which could be labeled as "craft beer drinkers". Today, this new beer segment constitutes the largest growing segment in the beer industry (Jaeger et al., 2020).

How is the craft beer consumer? What are the most valued craft beer attributes?. Craft beer consumption has been described as an experience delivering drink that offers pleasure, enjoyment, self-fulfillment, and social recognition to consumers (Berkhout et al., 2013; Aquilani et al., 2015; Gómez-Corona et al., 2016). Further, craft beer offers values that create differentiation in the actual beer market, whereby consumers are demanding a more unique and premium beer (Berkhout et al., 2013; Donadini et al., 2016). Prior studies indicate that beer consumers place great importance on the type of container, followed by the brewing technology and by the characteristics of raw materials used for brewing. Conversely, the least valued attributes are the brewery equipment and the brewery location (Donadini and Porretta, 2017).

It is important to remark that craft beer consumers behave differently from industrial beer drinkers. In the first place, the sensory stimulation produced by craft beer is differentiated from that produced by industrial beer in terms of flavor, aroma, texture, and color. In turn, craft beers are described as driving a higher sensory stimulation (Gómez-Corona et al., 2016). Accordingly, craft beer consumers perceive industrial beer as a secondary, less special product (Gómez-Corona et al., 2016).

In the second place, craft beer consumers could be characterized as being highly involved consumers, who have a higher general liking for beer (Jaeger et al., 2017a) and drink beer more often compared to industrial beer consumers (Aquilani et al., 2015). Due to their higher levels of involvement, these consumers desire to have higher knowledge about the product. So, the more information they have, the better they will enjoy craft beer as if knowledge and previous consumption experiences would enhance the drinking experience (Gómez-Corona et al., 2017c). These consumers also have greater involvement in beer and product-focused behaviors and activities (Jaeger et al., 2020).

In the third place, craft beer consumers could be described as being strongly motivated by the discovery of new flavors and the search for quality and product craftsmanship (Aquilani et al., 2015). Similarly, these consumers are generally differentiated from industrial beer drinkers in terms of their preference for innovative beers with novel and complex flavor profiles. So, it can be stated that craft beer consumers' liking is maximized when the beer contains an optimum amount of novelty (Giacalone, 2013) and that these consumers search for new and different beer products, and even search for a different lifestyle (Gómez-Corona et al., 2016).

Therefore, craft beer consumers can be described as being passionate about beer and demanding quality, variety, and novelty over mass-produced industrial beer alternatives.

How are the beer shopping and consumption process for craft beer consumers?

Craft beer consumers choose their beers with great care, showing a high need for unique beverages (Jaeger et al., 2020). As explained before, these consumers often want to know more about the beer they drink, from reading and taking notes of the beers they have already tasted (Gómez-Corona et al., 2016). Additionally, craft beer consumers develop a reflexive process when drinking since they usually take their time to think about the beer's flavor, and they often compare their sensations with other beer consumption previous experiences (Gómez-Corona et al., 2016).

Therefore, some authors, like Gómez-Corona et al. (2017b), highlight that craft beer consumption is characterized by the cognitive dimension of the consumption experience. Thus, this means that the consumers' relationship with craft beer relies on more on cognitive processes, compared to the consumption of industrial beer. More precisely, drinking craft beer is associated with thinking, memorizing, mentally representing the product, and creating meaning in consumption (Gómez-Corona et al., 2017b).

Regarding the consumption situation, craft beer consumers report that special beers are not that suitable to drink at parties or sports events since they like to drink these beers slowly (Silva et al., 2016). It noteworthy as craft beer is not considered an everyday product. More specifically, craft beer consumption moments are oriented towards special and ritualized moments, to drink alone to enjoy the product, with a close and small group of friends, or even on special occasions (Gómez-Corona et al., 2016), being craft beer the central element of consumption. Finally, and regarding the place where craft beer consumers buy this product, studies indicate that these consumers tend to shop at specialty beer stores (Gómez-Corona et al., 2016).

DRIVERS OF CRAFT BEER CONSUMPTION

Demand for Higher Quality

Consumers perceive craft beer as a high-quality product (Gómez-Corona et al., 2016), and with higher quality than industrial beer (Aquilani et al., 2015), but how is high quality perceived and what are the factors driving a high perceived quality?. Prior research highlights that the individual ability to perceive and detect quality depends on the familiarity with the specific product category. For this reason, most of the consumers who had drunk and tasted craft beer consider it to be of higher quality (Aquilani et al., 2015). Conversely, consumers who are unfamiliar with craft beer are unlikely to perceive the quality of a craft beer to be superior to that of industrial beers (Donadini and Porretta, 2017).

Most of the previous studies report that the high quality of craft beer is closely related to the materials and ingredients used for its production, as well as the brewing processes. In this context, authors like Aquilani et al. (2015) indicate that craft beer is perceived as with higher quality than an industrial beer due to the raw materials used for brewing. Similarly, Donadini and Porretta (2017) highlight that consumers indicate the origin of the raw materials used by craft brewers and the fact that craft beer is produced on a small scale as the significant characteristics of craft beer. Likewise, some authors note that (untraditional, local) ingredients and brewing processes are the most relevant factors and the most significant attributes of craft beer quality perception (Donadini et al., 2016; Koch and Sauerbronn, 2019).

Interestingly some extrinsic craft beer attributes that are present in the craft beer consumption experience also influence the high-quality perception of this alcoholic beverage. In the first place, the presence of sophisticated, attractive packaging and labels that can enhance the craft beer experience (Gómez-Corona et al., 2016). In the second place, the type of beer container is also a relevant attribute and a powerful influence on the craft beer quality judgment (Donadini and Porretta, 2017). More precisely, the container design, and its material, and the elements of beer packaging are of great importance for craft beer consumers. In this vein, glass is considered the preferred packaging material for craft beer, as well as a twist-off cap when a bottle is provided are highly appreciated by consumers.

The Search for Uniqueness

Following Mattiacci and Vignali (2004), unique products could be characterized as those that have distinctive or superior qualities that make them stand out from other products in their category. Often, this product distinctiveness and superior qualities are defined in terms of the objective properties of the product, its geographical origin, or even the cultural identity.

Regarding beer products, Cardello et al. (2016) indicate that unique products could be defined as "those that are highly differentiated from other products in their category, on the basis of perceivable sensory, image, functional, emotional or other product characteristics that are positively valued by consumers", suggest that uniqueness evokes positive connotations related to the desirability of the product. Additionally, Cardello et al. (2016) describe unique products as those with an unusual or complex character that provide novelty and excitement, having positive hedonic and emotional associations, and appropriateness many situational uses.

Besides, the need for uniqueness has been defined in the marketing literature as those individual differences among consumers in their propensity to desire goods services or experiences that few other consumers possess (Lynn and Harris, 1997). Similarly, from the marketing perspective, the concept of a unique product is defined as "one of a kind" or an "unusual" or "novel" product (Jaeger et al., 2017b).

Regarding beer consumption, previous studies highlight that beers that are identified as unique are those that are perceived as unfamiliar, unusual, highly differentiated, and novel. Further, consumers perceive craft beer as being unique and authentic, compared to industrial, commercial beer (Gómez-Corona et al., 2016). More specifically, the primary motivation for drinking craft beer seems to be the search for authenticity and the desire to find a particular identity in the product (Gómez-Corona et al., 2016). In this context, prior research describes craft beer consumers as having a strong desire for distinction, meaning that they do not want to consume what everyone else does (Gómez-Corona et al., 2017a), as well as having a desire for unique consumer products (Rivaroli et al., 2018; Rivaroli et al., 2019).

Following Cardello et al. (2016) craft unique beers add value to the consumer by their appeal to those consumers looking for something unusual or "special". For this reason, craft beer uniqueness reduces the likelihood that any other alcoholic beverage could serve as a suitable substitute for consumers (Cardello et al., 2016).

One of the reasons why craft beer is perceived as unique is the use of unusual ingredients for its production, the alternative methods to process ingredients, as well as the incorporation of local grain, herbs and spices (Schnell and Reese, 2003; Kleban and Nickerson, 2012; Reid et al., 2014).

In particular, unconventional ingredients are used by craft brewers to confer unique characteristics and souls to craft beer, which particularly valuable in signaling the quality of craft beer in consumers' minds (Donadini et al., 2016; Donadini and Porretta, 2017).

Furthermore, some authors, such as Donadini and Porretta (2017), report that the use of local ingredients is of great importance in defining the geographic proximity of craft beer. Local ingredients constitute a clear alternative to the global food model followed by large industrial scale brewers. This model often incorporates ingredients and raw materials, traveling long distances. Therefore, it can be stated that consumers recognize that craft beer is an expression of the agricultural identity of the territory - "terroir"- carrying geographical connotations (Donadini and Porretta, 2017).

Original Flavor and Exciting Taste

One of the most important trends today is the growing consumer interest in exploring new beer flavors and beer styles (Ascher, 2012), as well as the increasing demand for exciting, intriguing, and not dull beer flavors.

What are the main characteristics of beer taste?. Previous research reports that consumers expect to find flavors in beer such as bitterness, texture characteristics such as sparkles, as well as the quality of being a "thirst-quencher" when tasting beer (Sester et al., 2013).

Regarding craft beer, prior studies indicate that craft beer consumers enjoy the flavor, aroma, texture, and color of the beer, which could be described as complex, intense, and strong aromas, persistent flavors, bitterness, and dense texture (Donadini and Porretta, 2017). So, the craft beer sensory experience is triggered by the complex stimulation of this beer through flavor and taste (Gómez-Corona et al., 2017c). More precisely, Gómez-Corona et al. (2017c) suggest that flavor variety is one of the attributes highly appreciated by craft beer consumers that could be described as a "flavor paradise", compared to the flavor options perceived in the industrial beer. Similarly, craft beer consumers consider flavor as a characterizing element of craft beer (Donadini and Porretta, 2017). Finally, Sester et al. (2013) report that taste may be the dominating factor for beer choice.

On the other hand, some authors report that craft beer is chosen according to different flavor options, compared to commercial beer (Sester et al., 2013; Aquilani et al., 2015). Craft beer substantially differs from commercial, industrial beers, due to the traditional brewing methods are used for producing it, and for this reason, craft beers are valued for the high economic perception associated to them (Kleban and Nickerson, 2012).

What flavors are demanding craft beer consumers?. The exponential growth in the flavors and types of craft beers on the market suggests that craft beer consumers demand and enjoy a wide variety of beer flavor qualities, flavor intensities, and styles, also including highly innovative beers (Jaeger et al., 2020). Other authors indicate that craft beer consumers pay more attention to the aroma, foam, carbonation, and overall beer quality, compared to industrial beer consumers (Aquilani et al., 2015).

Finally, and regarding the different flavors of craft and industrial beer, it can be stated that many craft beer drinkers have come to eschew industrial beers with low flavor impact (Gómez-Corona et al., 2016). According to Donadini and Porretta (2017), consumers describe craft beer flavor as quite different from that of industrial beers, due to the use of unusual ingredients (Donadini and Porretta, 2017).

The Search for Status and Individuality

Following Elliott (1997), consumers consume symbolic goods to construct, express, and enhance their identity, being their primary consumption motivation to satisfy their need for social recognition. Accordingly, other authors like Witt (2010) note that the act of consumption serves as a way to signal social status and self-esteem, being this consumption mostly motivated by the symbolic value that products have.

Regarding the symbolic value of products, industrial, commercial beers could be considered as standardized products, or alcoholic filtered drinks transformed into a hardly identifiable object deprived of any identity. On the other hand, craft beer is consumed for its symbolic value (Gómez-Corona et al., 2016), instead of its functional attributes. Previous studies have ascribed the attributes of "uniqueness value" to premium craft beer, because of its connotation of "exclusivity" and "rareness".

Some authors indicate that one of the essential motives associated with craft beer consumption is social status, since, for some consumers, the interest in craft beer can signify status, similar to wearing a luxury branded product (Holt, 1995). So, craft beer consumers could be characterized as being status-seekers. For this reason, the consumption of craft beer is strongly associated with the communication of high social status (Dawson and Cavell, 1987). Some craft beer consumers believe that being identified as a "craft beer drinker" offers a higher perceived social status than the status achieved by being an "industrial beer drinker" (Jaeger et al., 2020). Therefore, one potential explanation for the higher social status ascribed to craft beer consumption is the desire for distinction since most craft beer consumers fell that they are not drinking what everyone else does (Gómez-Corona et al., 2017a).

Also, the consumption of craft beer is strongly associated with the desire to distinguish and making oneself different from others (Rivaroli et al., 2019). In this vein, authors like Choi and Stack (2005) report that beer consumers choose alternative specialty beers to express their individuality. More specifically, craft beer consumers want to differentiate themselves and express their individuality through their choice of purchasing and consuming alternative beers (Gómez-Corona et al., 2016).

Search of Self-Identity

The term self-identity could be defined as the individuals' use of a set of socially meaningful product categories to describe their identity and their role in the social setting (Rivaroli et al., 2019). Therefore, the search for self-identity by craft beer consumers may be grounded on the willingness to affirm their enduring self-perception and self-identity as craftwork-product consumers. In this context, some studies report that the search for self-identity may positively influence the behavior of consumers that engage in "craft consumption" as an act of self-expression (Campbell, 2005).

Some previous studies point out that the goal of craft beer consumption is not functional, but symbolic, as a desire to build self-identity and distinction (Gómez-Corona et al., 2016). So, it can be stated that craft beer consumers do not drink this product for its functional attributes, they consume it for what it means and what it represents as a more authentic and unique alcoholic beverage, compared to the mainstream industrial beer. Similarly, authors like Koch and Sauerbronn (2019) report craft beer consumers drink this product to maintain their self-concept and identity, through the development of an internalized emotional attachment to the product. This emotional attachment to the product may influence their enjoyment of beer and their effort to understand the product more deeply. The craft beer consumers' emotional attachment to this product is related to the product as perceived as something made with dedication and great affection, rather than as industrial merchandise (Koch and Sauerbronn, 2019).

In addition, it can be remarked that consumers perceive a significant difference between industrial beer and craft beer. On the one hand, industrial beer is recognized as a low-quality characterless beer, while on the other hand, craft beer is perceived as a beer made as a combination of unique ingredients praised brewers. Finally, craft beer consumers are described in the previous literature as having a strong self-identity (Rivaroli et al., 2018) and as consumers who do not want to be seen as drunks or related with any negative connotations associated with alcoholic drinking (Koch and Sauerbronn, 2019).

Willingness to Pay a Premium Price

Retail price is a factor of particular consumer relevance in the beer sector. In this vein, previous studies report that craft beer is an alcoholic beverage appealing for consumers who are willing to pay more for a beer (Gomez-Corona et al., 2016), and craft beer consumers often avoid mass-produced beer and are not influenced by price. Further, some consumers show a great willingness to pay higher prices for high quality differentiated beers (Ascher, 2012). Similarly, craft beer is an alcoholic product that should not be perceived as a cheap alcoholic drink to get drunk (Koch and Sauerbronn, 2019).

However, the price of craft beer is also considered as one crucial barrier for craft beer consumption, since consumers always compare craft beer price with that of commercial, industrial beers (Gómez-Corona et al., 2016). More precisely, for some consumers, the price of craft beer is reasonable, while for others, the price of craft beer indicates that this is just a product for special occasions.

Opposite to craft beer, industrially brewed beers are perceived as convenient utilitarian products, being cheap affordable drinks easily accessible for consumers. Likewise, industrial beer consumers are used to drinking mass-produced beer usually priced in the bottom and mostly under the price of 2.00 euros. For this reason, they do not expect that much of the quality offered by craft beers (Donadini and Porretta, 2017).

It is important to remark the direct relationship between price and perceived product quality. Regarding beer, higher prices are associated with higher levels of beer quality. In comparison, lower or cheap prices are often associated with alternatively lower quality, unpleasant taste, and more affordable mass-produced beer (Ascher, 2012).

Accordingly, unique and special alcoholic beverages that could be characterized as having superior quality enable setting higher prices (Mattiacci and Vignali, 2004). In the context of beer consumption, beer consumers describe the mass-produced industrial beer as being a cheap and affordable drink; while considering craft beers as having generally higher prices (Ascher, 2012). However, beer consumers perceive that the high quality of craft beer enables paying less attention to price (Aquilani et al., 2015).

CRAFT BEER DISTRIBUTION

On-Trade Distribution (Breweries, Pubs …)

Craft beer is not perceived as an "everyday product", and for this reason, its distribution is different from industrial beer.

Prior research indicates that craft beer is mainly consumed in brewpubs, bars, cafeterias, and restaurants. What aforementioned confirms that craft beer is mostly consumed in places traditionally connected to its consumption and showing that its consumptions have not been substituted by off-trade use (Aquilani et al., 2015). There is one potential explanation for this on-trade consumption. According to Keeling Bond et al. (2009), craft beer consumers decide to buy craft beer directly from breweries appreciating not only location-specific attributes of beer, but also the great variety of product characteristics, the production process, the shopping convenience, and some traditional product-specific traits.

Also, some authors aim to explain why craft beer is hugely available in the marketplace. One possible explanation is the existence of an increasing number of microbreweries worldwide. Another potential reason is because of the great desire of craft breweries to provide innovative beers (Jaeger et al., 2020).

Table 2. Factors influencing the consumption of craft beer

Factor	Type of Influence	Previous Research
Higher Quality	Craft beer perception as a high-quality product	Aquilani et al. (2015) Gómez-Corona et al. (2016)
	Small-scale production, brewing processes and raw materials and ingredients used provides high quality to craft beer	Aquilani et al. (2015) Donadini & Porretta (2017) Koch & Sauerbronn (2019)
	The type of beer container, design of the container, materials, and elements of packaging influence quality perception	Donadini & Porretta (2017) Gómez-Corona et al. (2016)
	Glass as packing material and the twist-off cap is preferred to infer beer high quality	Aquilani et al. (2015).
Uniqueness	Craft beer has uniqueness value, with connotations of exclusivity and rareness	Dawson & Cavell (1987)
	Uniqueness and authenticity are two critical motivations for drinking craft beer	Gómez-Corona et al. (2016) Rivaroli et al. (2018) Rivaroli et al. (2019)
	Craft beer is perceived as unique and is also considered unfamiliar, unusual, highly differentiated and with special sensory properties	Cardello et al. (2016)
Uniqueness	Craft beer uniqueness derives from the use of unconventional ingredients and alternative brewing processes	Schnell & Reese (2003) Kleban & Nickerson (2012) Reid et al. (2014) Donadini et al. (2016)
	Craft beer as an expression of "terroir" identity, with geographical connotations	Donadini & Porretta (2017)
	Craft breweries differentiate themselves using unique local ingredients that are abundant in the region or "terroir."	Gómez-Corona et al. (2016)
Flavor and Taste	The flavor of craft beer is described as rich, complex, intense and composite	Donadini & Porretta (2017)
	Craft beer is consumed for its variety of flavors	Sester et al. (2013) Aquilani et al. (2015)
	Craft beer consumers pay more attention to flavor and aroma than industrial beer drinkers	Aquilani et al. (2015) Gómez-Corona et al. (2017c)
	Craft beer consumers demand a wide variety of flavors, as well as innovative flavors	Jaeger et al. (2020)

Table 2. (Continued)

Social Status	Craft beer consumers are characterized as seeking social status and distinguish themselves from other consumers	Holt (1995) Rivaroli et al. (2019)
	Craft beer consumers want to move away from industrial beer consumption	Gómez-Corona et al. (2016)
	The consumption of craft beer is strongly associated with a high social status	Dawson & Cavell (1987) Gómez-Corona et al. (2017a)
Individuality	One motivation to consume craft beer is to express individuality	Choi & Stack (2005)
Self-Identity	Craft beer consumption is grounded on consumers' need to affirm their self-perception as craftwork product consumers	Rivaroli et al. (2019)
	Craft beer consumers are characterized as having strong self-identity or as having a desire to build their own identity	Gómez-Corona et al. (2016) Rivaroli et al. (2018)
	A desire for self-expression drives craft beer consumers	Campbell (2005)
	Consumers drink craft beer to maintain their self-concept and identity	Koch & Sauerbronn (2019)
Premium Price	Higher prices are associated with higher levels of beer quality	Ascher (2012) Mattiacci & Vignali (2004)
	Craft beer appeals to the consumer who wants to pay premium prices for specialty	Gómez-Corona et al. (2016)
	Premium prices are considered a barrier for craft beer consumption for some consumers	Gómez-Corona et al. (2016)
	Some craft beer consumers are not influenced by price and avoid mass-produced industrial beer	Koch & Sauerbronn (2019)
On-Trade Consumption	Consumers shop on-trade due to their internalization of beer location-specific attributes	Keeling Bond et al. (2009)
	Craft beer is mostly consumed in places traditionally connected to its consumption	Aquilani et al. (2015)
	Some consumers perceive shopping directly from brewers as inconvenient	Donadini & Porretta (2017)
	One characteristic of craft beer is the shortness of the supply chain	Donadini & Porretta (2017)
Off-Trade Consumption	The availability of craft beer in off-trade channels is perceived as a barrier for some consumers	Gómez-Corona et al. (2016) Donadini & Porretta (2017)
	For some consumers shopping craft beer in specialty stores is part of the experience of craft beer consumption	Gómez-Corona et al. (2016) Jaeger et al. (2020)

Source: Own elaboration.

Retail or Off-Trade Distribution

In general terms, it can be stated that craft beer is not an alcoholic beverage or a drink consumed off-trade since consumers have a high propensity to seek craft beers from small local breweries (Jaeger et al., 2020).

However, the low availability of craft beer in the mainstream food retailing system is considered as a barrier for some consumers, who wish to buy craft beer everywhere, such for example in traditional supermarkets or convenience stores (Gómez-Corona et al., 2016). Some authors note that consumers perceive the nature of the retail outlet where craft beer is sold to have low availability and accessibility (Donadini and Porretta, 2017). As a consequence, craft beer is not perceived as to be easily accessible or largely available to consumers.

Furthermore, most of the consumers perceive shopping directly from brewers as inconvenient, given that craft breweries are not seen as businesses located close to shoppers. Conversely, authors like Donadini and Porretta (2017) indicate that the shortness of the supply chain is an element of distinction of craft beer that is perceived by consumers as a measure of the geographical distance between the brewer and the suppliers of ingredients and raw materials.

On the other hand, consumers perceive that craft beer is mostly sold in beer shops and breweries, rather than in supermarket or hypermarket chains. Interestingly, for some consumers, the act of going out and looking for craft beer in small, specialized stores is part of the exciting craft beer experience (Gómez-Corona et al., 2016).

CONCLUSION

Nowadays, consumer beer preferences seem to be related to the search for higher product quality and craftsmanship. As a consequence, beer consumers are more inclined to drink craft beer, which has different attributes and characteristics compared to the mass-marketed industrial beer.

These factors are altering the beer consumption habits, preferences, and demand, representing a signal of a growing beer culture that is spreading worldwide. This emerging beer culture could be labeled as the "craft beer revolution".

This phenomenon is explained and developed in this chapter, also examining the main factors that drive the consumption of this alcoholic beverage.

Our findings suggest that considering the low switching costs of mass-produced beer to alternative alcoholic beverages, such as wine or spirits, brewing companies should consider the production of craft beer as an alternative option with high added-value.

This research can help large manufacturing brewing companies to understand the new consumption trend better, labeled as "the revolution of craft beer", and to meet the needs, wants, and preferences of beer consumers today to produce the best possible craft beer. Further, the study presented here could provide useful insights to brewing managers to promote their beer brands by changing their characteristics or communication and distribution channels to meet the new consumer expectations and preferences about beer.

REFERENCES

Acitelli, Tom, and Tonny Magge. 2017. *"The audacity of hops: The history of America's craft beer revolution."* Chicago: Chicago Review Press.

American Brewers Association. 2016. *"Craft brewer defined"*. http://www.brewersassociation.org/pages/businesstools/craft-brewing-statistics/craft-brewer-defined.

American Brewers Association. 2020. *"Craft Brewer Definition."* https://www.brewersassociation.org/statistics-and-data/craft-brewer-definition/.

Aquilani, Barbara, Tiziana Laureti, Stefano Poponi, and Luca Secondi. 2015. "Beer choice and consumption determinants when craft beers are tasted: An exploratory study of consumer preferences." *Food Quality and Preference* 41: 214–224.

Argent, Neil. 2018. "Heading down to the Local? Australian Rural Development and the Evolving Spatiality of the Craft Beer Sector." *Journal of Rural Studies* 61: 84–99.

Ascher, Bernard. 2012. *"Global beer: The road to monopoly"*. American Antitrust Institute. Retrieved from http://www.antitrustinstitute.org/content/global-beer-roadmonopoly-.

AssoBirra. 2014. *"Annual report 2013"*. Retrieved from http://www.assobirra.it/press/ wp-content/ar2013-assobirra_03-09.pdf

AssoBirra. 2016. *"Annual report 2016"*. Retrieved from http://www.assobirra.it/wpcontent/uploads/2017/06/ASSOBIRRA_AR_2016.pdf

AssoBirra. 2020. *"Annual Report 2020"*. Retrieved from . https://www.assobirra.it/pubblicazioni/.

Bentzen, Jan, and Valdemar Smith. 2018. "Entry, Survival, and Profits: The Emergence of Microbreweries in Denmark." In *Economic Perspectives on Craft Beer*, edited by Christian Garavaglia and Johan Swinnen, 161–82. Springer Nature.

Berkhout, Bram, Lianne Bertling, Yannick Bleeker, Walter de Wit, Geerten Kruis, Robin Stokkel, and Ri-janne Theuws. 2013. *"The contribution made by beer to the European economy."* A report commissioned by The Brewers of Europe and conducted by Regioplan Policy Research and EY.

Cabras, Ignazio, and Charles Bamforth. 2016. "From Reviving Tradition to Fostering Innovation and Changing Marketing: The Evolution of Micro-Brewing in the UK and US, 1980–2012." *Business History* 58 (5): 625–46.

Campbell, Colin. 2005. "The craft consumer." *Journal of Consumer Culture* 5(1): 23–42.

Cannatelli, Benedetto, and Matteo Pedrini. 2012. *"Osservatorio ALTIS-UNIONBIRRAI Sul Segmento Della Birra Artigianale in Italia - Rapporto 2011."* ["*ALTIS-UNIONBIRRAI Observatory on the Craft Beer Segment in Italy - 2011 Report.*"] Milano.

Cannatelli, Benedetto, Matteo Pedrini, and Marco Grumo. 2015. "The effect of brand management product quality on firm performance." *Journal of Food Products Marketing* 2: 1–24.

Cardello, Armand V., Benedicte Pineau, Amy G. Paisley, Christina M. Roigard, Sok L. Chheang, Lim F. Guo, and Sara R. Jaeger. 2016. "Cognitive and emotional differentiators for beer: An exploratory study focusing on "uniqueness." *Food Quality and Preference*, 54: 23–38.

Charters, Steve, and Simone Pettigrew. 2008. "Why do people drink wine? A consumer-focused exploration." *Journal of Food Products Marketing* 14(3): 13–32.

Choi, David Y., and Martin H. Stack. 2005. "The all-American beer: A case of inferior standard (taste) prevailing?." *Business Horizons* 48(1):78-89.

Colen, Liesbeth, and Johan F. Swinnen. 2011. "Beer Drinking Nations: The Determinants of GlobalBeer Consumption." In J.F.M. Swinnen (ed.). *The Economics of Beer*. Oxford: Oxford University Press.

CRAFT. 2019. "*Definitie CRAFT-Brouwer*." https://craftbrouwers.nl/organisatie/definitie/.

Danson, Mike, Laura Galloway, Ignazio Cabras, and Tina Beatty. 2015. "Microbrewing and Entrepreneurship: The Origins, Development and Integration of Real Ale Breweries in the UK." *International Journal of Entrepreneurship and Innovation* 16 (2): 135–44.

Dawson, Scott, and Jill Cavell. 1987. "Status recognition in the 1980s: Invidious distinction revisited." *ACR North American Advances* 14: 487–491.

Donadini, Gianluca, Maria D. Fumi, Edyta Kordialik-Bogacka, Launa Maggi, Milena Lambri, and Paolo Sckokai. 2016. "Consumer interest in Specialty Beers in three European markets." *Food Research International* 85: 301–314.

Donadini, Gianluca, and Sebastiano Porretta. 2017. "Uncovering patterns of consumers' interest for beer: A case study with craft beers." *Food Research International* 91: 183–198.

Elliott, Richard. 1997. "Existential consumption and irrational desire." *European Journal of Marketing* 31: 285–296.

Elzinga, Kenneth G., Carol Horton Tremblay, and Victor J. Tremblay. 2015. "Craft Beer in the United States: History, Numbers, and Geography." *Journal of Wine Economics* 10(3): 242–74.

Euromalt. 2020. "*Euromalt Statistics - Malt Production Capacity 2017.*" http://www.euromalt.be/data/15401964073.- Malt Production Capacity 2017.pdf.

Fastigi, Matteo, Elena Viganò, and Roberto Esposti. 2018. "The Italian Microbrewing Experience: Features and Perspectives." *Bio-Based and Applied Economics* 7 (1): 59–86.

Garavaglia, Christian, and David Castro. 2018. "The Recent Advent of Micro Producers in the Spanish Brewing Industry." In *Economic Perspectives on Craft Beer*, edited by Christian Garavaglia and Johan Swinnen, 345–72. Springer Nature.

Garavaglia, Christian, and Johan Swinnen. 2018a. "Economic of the Craft Beer Revolution: A Comparative International Perspective." In *Economic Perspectives on Craft Beer*, edited by Christian Garavaglia and Johan Swinnen, 3–51. Cham, Switzerland: Springer Nature.

———. 2018b. *Economic Perspectives on Craft Beer - A Revolution in the Global Beer Industry*. Edited by Christian Garavaglia and Johan Swinnen. *Economic Perspectives on Craft Beer*. Cham, Switzerland: Springer Nature.

Giacalone, Davide. 2013. "*Consumers' perception of novel beers: Sensory, affective, and cognitive contextual aspects*". University of Copenhagen: Copenhagen, Denmark.

Gómez-Corona, Carlos, Héctor B. Escalona-Buendía, García, M., Sylvie Chollet, and Dominique Valentin. 2016. "Craft vs. industrial: Habits, attitudes and motivations towards beer consumption in Mexico." *Appetite* 96: 358–367.

Gómez-Corona, Carlos, Dominique Valentin, Héctor B. Escalona-Buendía, and Sylvie Chollet. 2017a. "The role of gender and product consumption in the mental representation of industrial and craft beers: An exploratory study with Mexican consumers." *Food Quality & Preference* 60: 31-39.

Gómez-Corona, Carlos, Sylvie Chollet, Héctor B. Escalona-Buendía, and Dominique Valentin. 2017b. "Measuring the drinking experience of beer in real context situations. The impact of affects, senses, and cognition." *Food Quality & Preference* 60: 113-122.

Gómez-Corona, Carlos, Héctor B. Escalona-Buendía, Sylvie Chollet, and Dominique Valentin. 2017c. "The building blocks of drinking experience across men and women: A case study with craft and industrial beers." *Appetite* 116: 345-356.

Gracia-Arnaiz, Mabel. 2011. "Culture, Market and Beer Consumption". In Schiefenhovel, W., & McBeth, H. (eds.) *Liquid Bread Beer and Brewing in Cross-Cultural.* Oxford UK: Berghahn Books.

Gutjar, Swetlana, Jelle R. Dalenberg, Cees de Graaf, René A. de Wijk, Aikaterini Palascha, Remco J. Renken, and Gerry Jager. 2015. "What reported food-evoked emotions may add: A model to predict consumer food choice." *Food Quality & Preference* 45: 140–148.

Jaeger, Sara R., Armand V. Cardello, Sol K. Chheang, Michelle K. Beresford, Duncan I. Hedderley, and Benedicte Pineau. 2017a. "Holistic and consumer-centric assessment of beer: A multi-measurement approach." *Food Research International* 99: 287-297.

Jaeger, Sara R., Armand V. Cardello, David Jin, Denise C. Hunter, Christina M Roigard, and Duncan I. Hedderley. 2017b. "Product uniqueness: Further exploration and application of a consumer based methodology." *Product Quality & Preference* 60: 59-71.

Jaeger, Sara R., Thierry Worch, Tracey Phelps, David Jin, and Armand V. Cardello. 2020. "Preference segments among declared craft beer drinkers: Perceptual, attitudinal and behavioral responses underlying craft-style vs. traditional-style flavor preferences." *Food Quality & Preference* 82: 103884.

Keeling Bond, Jennifer, Dawn Thilmany, and Craig Bond. 2009. "What influences consumer choice of fresh produce purchase location?." *Journal of Agricultural and Applied Economics* 41(1): 61–74.

Kiefer, David M. 2001. "*Brewing: A legacy of ancient times. Today's chemist at work*" (41– 42). American Chemical Society.

Kleban, Jack, and Inge Nickerson. 2012. "To brew, or not to brew-that is the question: An analysis of competitive forces in the craft brew industry." *Journal of the International Academy for Case Studies* 18(3): 59-82.

Koch, Eduardo S., and Joao F. Sauerbronn. 2019. "To love beer above all things": An analysis of Brazilian craft beer subculture of consumption." *Journal of Food Products Marketing,* 25(1): 1-25.

Law No. 154/2016. n.d. *Gazzetta Ufficiale Della Repubblica Italiana.* https://www.gazzettaufficiale.it/eli/id/2016/08/10/16G00169/sg.

Lynn, Michael, and Judy Harris. 1997. "The desire for unique consumer products: A new individual differences scale." *Psychology & Marketing* 14(6): 601–616.

Mattiacci, Alberto, and Claudio Vignali. 2004. "The typical products within food "glocalisation": The makings of a twenty-first-century industry." *British Food Journal* 106(10-11): 703–713.

Menghini, Silvio, Veronica Alampi Sottini, Bruno Fabbri, and Sara Fabrizzi. 2019. *"Report 2018 Birra Artigianale Filiera Italiana e Mercati."* [*"Report 2018 Craft Beer Italian Supply Chain and Markets."*] Firenze. https://www.unionbirrai.it/it/news/report-ub-obiart-2018-birra-artigianale-filiera-italiana-e-mercato/.

Pettigrew, Simone, and Steve Charters. 2006. "Consumers' expectations of food and alcohol pairing." *British Food Journal* 108(3): 169–180.

Ravelli, Giovanni, and Matteo Pedrini. 2015. *"Osservatorio ALTIS – UNIONBIRRAI Sul Segmento Della Birra Artigianale in Italia (Rapporto 2015)."* [*"ALTIS - UNIONBIRRAI Observatory on the Craft Beer Segment in Italy (2015 Report)."*] Milano. https://altis.unicatt.it/altis-Altis_UB_2012. pdf.

Reid, Neil, Ralph B. McLaughlin, and Michaes S. Moore. 2014. "From Yellow Fizz to Big Biz: American Craft Beer Comes of Age." *Focus on Geography* 57(3): 114-125.

Rivaroli, Sergio, Martin K. Hingley, and Roberta Spadoni. 2018. "The motivation behind drinking craft beer in Italian brew pubs: A case study." *Economia Agro-Alimentare* 20(3): 425–443.

Rivaroli, Sergio, Jörg Lindenmeier, and Roberta Spadoni. 2019. "Attitudes and Motivations Toward Craft Beer Consumption: An Explanatory Study in Two Different Countries." *Journal of Food Products Marketing* 25(3): 276-294.

Schnell, Steven M., and Joseph F. Reese. 2003. "Microbreweries as tools of local identity." *Journal of Cultural Geography* 21(1): 45–69.

Sester Carole, Ophelia Deroy, Angela Sutan, Fabrice Galia, Jean-François Desmarchelier, and Dominique Valentin. 2013. "Having a drink in a bar": an immersive approach to explore the effects of context on drink choice." *Food Quality & Preference* 28: 23–31.

Silva, Ana P., Gerry Jager, Roelien van Bommel, Hannelize van Zyl Hans-Peter Voss, Tim Hogg, Manuela Pintado, and Cees de Graaf. 2016. "Functional or emotional? How Dutch and Portuguese conceptualise beer, wine and non-alcoholic beer consumption." *Food Quality & Preference* 49: 54-65.

The Brewers of Europe. 2019. "*European Beer Trends - Statistics Report / 2019 Edition.*" Brussels. https://brewersofeurope.org/uploads/mycms-files/documents/publications/2019/european-beer-trends-2019-web.pdf.

World Health Organization, WHO. 2018. Global Status Report on Alcohol and Health 2018. Edited by Vladimir Poznyak and Dag Rekve. *Global Status Report on Alcohol.* https://doi.org/10.1037/cou0000248.

Witt, Ulrich. 2010. "Symbolic consumption and the social construction of product characteristics." *Structural Change and Economic Dynamics* 21: 17–25.

BIOGRAPHICAL SKETCHES

Cristina Calvo-Porral, PhD

Affiliation: University of A Coruña, Spain

Education: MBA in Business Administration at University Pontificia de Comillas (ICADE, Madrid); PhD in Economics at University of A Coruña.

Business Address: Facultad de Economía y Empresa, Dpto. de Empresa. Campus de Elviña s/n, A Coruña-15071

Research and Professional Experience:

Cristina Calvo-Porral is Ph.D Full Marketing Professor at Business Department in University of La Coruña (SPAIN). She has been teaching as member of the faculty since year 2006.

Some of the courses given by Cristina Calvo-Porral are "Introduction to Marketing", "Market Research techniques", "Fundamentals of Marketing", "Marketing Strategies" and "Commercial Management".

She has directed two doctoral dissertations in year 2015 and year 2018.

Cristina Calvo-Porral has made numerous postdoctoral research stays at the Universite du Quebec en Outaouais (Ottawa, Canada) in years 2012, 2014 and 2017; St. Joseph's University (Philadelphia, USA) in year 2013; at the University of Texas at Austin (Austin, USA) in year 2015, and at the Universidad de la Frontera (Chile) in year 2018. She has been appointed as Research Professor at the Universite du Quebec en Outaouis (Ottawa, Canada) from year 2014.

Cristina Calvo-Porral has published 25 academic articles indexed in Web of Science, and 43 academic journals indexed in Scopus. She has authored numerous book chapters in the Marketing field.

Regarding her company experience, Cristina Calvo-Porral has professional experience as responsible at the Internationalization and Exports Department of the leading fashion company Carolina Herrera (Sociedad Textil Lonia, S.A.) from years 2002 to 2011.

Professional Appointments:

Full Marketing Professor at the University of A Coruña (SPAIN).

Publications from the Last 3 Years:

Year 2019:
1. Calvo-Porral, C., and Nieto Mengotti, M. V. (2019), "The moderating influence of involvement with ICTs in mobile services", *Spanish Journal of Marketing*, Vol.: 23(1), pp. 25-43.
2. Calvo-Porral, C., and Lévy Mangin, J. P. (2019), "Profiling Shooping Mall customers during hard times", *Journal of Retailing & Consumer Services*, Vol.: 49, pp.238-246
3. Calvo-Porral, C., Ruiz-Vega, A., and Lévy Mangin, J. P. (2019), "How consumer involvement influences consumption elicited emotions", *International Journal of Market Research,*
4. Calvo-Porral, C., and Lévy Mangin, J. P. (2019), "Situational factors in alcoholic beverage consumption: Examining the influence of the place of consumption", *British Food Journal*, Vol.: 121(9) pp. 2086-2101.
5. Calvo-Porral, C., Lévy Mangin, J. P., and Ruiz-Vega, A. (2019), "An emotion-based typology of wine consumers", *Food Quality & Preference*, Vol.: 79, 103777.
6. Calvo-Porral, C., and Pesqueira Sánchez, R. (2019), "Generational differences in technology behaviour: Comparing Millennials and Generation X", *Kybernetes* (*In Press*).

Year 2018:
1. Calvo-Porral, C., Orosa-González, J., and Blazquez Lozano, F. (2018), "A clustered-based segmentation of beer consumers: from "beer lovers" to "beer to fuddle", *British Food Journal*, Vol. 120(6),pp. 1280-1294.
2. Calvo-Porral, C., and Lévy Mangin, J. P. (2018), "Pull Factors of the Shopping Malls: An Empiciral Study", *International Journal of Retail & Distribution Management,* Vol. 46(2), pp. 110-124.

3. Calvo-Porral, C., Pesqueira Sánchez, R., and Faiña, J. A. (2018), "A Clustered-based Categorization of Millennials in their Technology Behavior", *International Journal of Human-Computer Interaction*, Vol.: 35(3), pp. 231-239.
4. Calvo-Porral, C., and Lévy Mangin, J. P. (2018), "From "foodies" to "cherry-pickers": A clustered-based segmentation of specialty food retail customers", *Journal of Retailing & Consumer Services*, Vol.: 43, pp. 278-284.
5. Calvo-Porral, C., Ruiz-Vega, A., and Lévy Mangin, J. P. (2018), "The influence of consumer involvement in wine consumption-elicited emotions", *Journal of International Food & Agribusiness Marketing,* Vol.: 31(2), pp. 128-149.
6. Calvo-Porral, C., Ruiz-Vega, A., and Lévy Mangin, J. P. (2018), "Does product involvement influence how emotions drive satisfaction?: An approach through the Theory of Hedonic Asymmetry", *European Research on Management & Business Economics,* Vol.: 24(3), pp. 130-136.

Year 2017:
1. Calvo-Porral, C., Faiña, J. A., and Losada-Lopez, C. (2017), "Can Marketing help in tackling Food Waste?: Proposals in Developed Countries", *Journal of Food Products Marketing*, Vol.: 23 (1), pp. 42-60 .
2. Calvo-Porral, C., and Lévy Mangin, J. P. (2017), "Store Brands' Purchase Intention: Examining the role of Perceived Quality", *European Research on Management & Business Economics,* Vol.: 23 (2), pp. 90-95.
3. Calvo-Porral, C., and Lévy Mangin, J. P. (2017), "Specialty Food Retailing: Examining the role of Products' Perceived Quality", *British Food Journal*, Vol.: 119(7), pp. 1511-1524 .

Sergio Rivaroli, PhD

Affiliation: Department of Agricultural and Food Sciences - Alma Mater Studiorum – Università di Bologna

Education: Agricultural Economics and Policy

Business Address: Viale Fanin G., 50 - 40127 Bologna (IT)

Research and Professional Experience: Consumer economics, food economics, and prosocial behavior.

Publications from the Last 3 Years:
1. Calvo-Porral, Cristina, Sergio Rivaroli, and Javier Orosa-González. 2020. "How Consumer Involvement Influences Beer Flavour Preferences." *International Journal of Wine Business Research.* https://doi.org/10.1108/IJWBR-10-2019-0054.
2. Rivaroli, Sergio, Beatrice Baldi, and Roberta Spadoni. 2020. "Consumers' Perception of Food Product Craftsmanship: A Review of Evidence." *Food Quality and Preference* 79 (May 2019): 103796. https://doi.org/10.1016/j.foodqual.2019.103796.
3. Rivaroli, Sergio, Jörg Lindenmeier, and Roberta Spadoni. 2020. "Is Craft Beer Consumption Genderless? Exploratory Evidence from Italy and Germany." *British Food Journal* 122 (3): 929–43. https://doi.org/10.1108/BFJ-06-2019-0429.
4. Spadoni, Roberta, Mattia Nanetti, Antonio Bondanese, and Sergio Rivaroli. 2019. "Innovative Solutions for the Wine Sector: The Role of Startups." *Wine Economics and Policy* 8 (2): 165–70. https://doi.org/10.1016/j.wep.2019.08.001.
5. Rivaroli, Sergio, Vratislav Kozák, and Roberta Spadoni. 2019. "What Motivates Czech and International 'Millennial-Aged' University Students to Consume Craft Beers?" *International Journal of Wine Business Research* 31 (3): 441–55. https://doi.org/10.1108/IJWBR-11-2018-0067.

6. Rivaroli, Sergio, Arianna Ruggeri, and Roberta Spadoni. 2019. "Food 'Buycott' as an Ethical Choice against Mafia in Italy." *Journal of Social Marketing* 9 (4): 490–506. https://doi.org/10.1108/JSOCM-11-2018-0139.
7. Rivaroli, Sergio, Jörg Lindenmeier, and Roberta Spadoni. 2019. "Attitudes and Motivations Toward Craft Beer Consumption: An Explanatory Study in Two Different Countries." *Journal of Food Products Marketing* 25 (3): 276–94. https://doi.org/10.1080/10454446.2018.1531802.
8. Rivaroli, Sergio, Arianna Ruggeri, Pietro Novi, and Roberta Spadoni. 2018. "Purchasing Food to Counteract Mafia in Italy." *Journal of Social Marketing* 8 (2): 142–58. https://doi.org/10.1108/JSOCM-03-2017-0019.
9. Scocco, Paola, Sergio Rivaroli, Francesca Mercati, Federico M. Tardella, Alessandro Malfatti, Elena De Felice, and Andrea Catorci. 2018. "Anatomy for Economy: Starting from the Rumen Keratinization Degree to Enhance the Farm Income." *Economia Agro-Alimentare* 20 (2): 261–72. https://doi.org/10.3280/ECAG2018-002010.
10. Rivaroli, Sergio, Martin K. Hingley, and Roberta Spadoni. 2018. "The Motivation behind Drinking Craft Beer in Italian Brew Pubs: A Case Study." *Economia Agro-Alimentare* 20 (3): 425–43. https://doi.org/10.3280/ECAG2018-003009.

In: Beer: From Production to Distribution ISBN: 978-1-53618-414-3
Editor: Armand Legault © 2020 Nova Science Publishers, Inc.

Chapter 3

EFFECT OF PROCESSING ON THE ANTIOXIDANT ACTIVITY OF BEER

*Iuliana Aprodu**
Faculty of Food Science and Engineering, "Dunarea de Jos" University of Galati, Romania

ABSTRACT

Beer is rich in bioactive compounds, mainly antioxidants, which provide health related benefits in case of moderate consumption. The antioxidant activity of beer is mainly due to the endogenous phenolic compounds with high bioavailability, Maillard reaction products and sulfites. The phenolics coming from malt account for up to 80% of the total phenolic content of beer and include phenolic acids and derivatives, anthocyanins, proanthocyanidins, lignans and lignin-related compounds. Additional phenolic acids, chalcones, flavonoids, catechins, and proanthocyanidins are derived from hops. In addition to the contribution to the formation of beer color, aroma and flavor, the final products of the Maillard reaction, melanoidins, exhibit strong antioxidant properties and other biological effects. Finally, the sulfites found in beer are fermentation by-products produced by yeasts while synthesizing of sulfur-containing

---

* Corresponding Author's Email: iuliana.aprodu@ugal.ro.

amino acids. The amount of endogenous antioxidants varies with the beer type, being affected by genetic and agricultural factors in the raw material and by processing during malting and brewing. The technological steps influence in different ways the antioxidant activity of beer. The amount of phenolic compounds and the overall antioxidant activity are significantly affected during malting, mashing, fermentation and storage. Better antioxidant activity of beer and improved retention of the biologically active compounds can be obtained by selecting the appropriate raw materials and optimizing the brewing parameters.

Keywords: beer, antioxidant activity, phenolic compounds, malting, brewing

INTRODUCTION

The main ingredients used for manufacturing beer are water, malt, non-malted cereals, hops and yeast. Beer is a complex beverage which contains about 800 organic compounds, in addition to water, ethanol and CO_2 (Buiatti, 2009). Beer composition highly depends on the quality of these raw materials, and on the fermentation by-products produced by yeasts, which are usually responsible for the particular flavor which makes the product unique.

The nutritional value of beer is higher compared to other alcoholic beverages (Tafulo et al., 2010). Some of the compounds found in beer originate from the raw materials and others are formed during fermentation, as a result of the metabolic activity of the yeast. Among the compounds originating from the raw materials, some are not effected during brewing, while others suffer changes during processing.

Malt is the main source of carbohydrate, protein, lipid and polyphenolic compounds in beer, while additional fermentable carbohydrates might arise from adjuncts. Anyway, these compounds suffer important changes during processing for beer manufacturing. For instance, protein and starch degradation is initiated during barley germination and is continued during mashing. The efficiency of these changes exert a high influence on the fermentation process and development of the flavor compounds found in beer. The use of malt and cereals for beer manufacturing allows obtaining products rich in vitamins.

The inorganic salts originating from water and/or malt may be found as such in the finale products, but most probably are precipitated in the brake in the brewing process or used by yeast during fermentation. Therefore, the spectrum of these compounds in the beer might differ significantly in respect to that of the raw materials (Buiatti, 2009). Beer was reported to be rich in cations like magnesium, potassium, sodium, and calcium and anions like chloride, sulfate, nitrate and phosphate.

Beer is an important source of bioantioxidants, which are dietary compounds which ensure important reduction of the damages to the normal physiological functions in humans caused by the reactive oxygen species (Zhou and Liu, 2005). The naturally occurring antioxidant compounds from beer originate mainly from barley and hops, the levels depending on the varieties (Agu, 2002; Derdelinckx, 2008; Liu and Yao, 2007). Therefore, the antioxidant activity of beer highly depends on the initial antioxidant content of these natural ingredients used in brewing and on the changes occurred during processing (Dabina-Bika et al., 2013). Moreover, the changes suffered by the bioactive compounds with antioxidant properties during processing, especially during malting, mashing, sparing, boiling and yeast fermentation should be also considered (Preedy and Rajendram, 2009).

The major antioxidant compounds found in beer are phenolics and Maillard compounds formed mainly during malting process and wort boiling (Goupy et al., 1999). Different research groups investigated the correlation between the total phenol content and antioxidant properties of beer (Piazzon et al., 2010; Zhao et al., 2010). In addition to the important contribution to the antioxidant capacity of beer, the phenolic compounds influence as well the colour, flavour, astringency and bitterness of the products, resulting in mouth-feel differences (Dabina-Bicka et al., 2013). The phenolic compounds influence beer stability, through their participation in chill haze formation (Goupy et al., 1999). The melanoidins are found in high concentrations in dark beer, influencing the colour, aroma and flavor of the product (Preedy and Rajendram, 2009; Dabina-Bicka et al., 2013).

The antioxidant capacity of beer is not exclusively due to the individual bioactive compounds originated from the raw materials or produced during brewing, but also to the synergistic effects between them, presence of trace elements, metals, and other constituents (Preedy and Rajendram, 2009; Dabina-Bicka et al., 2013). Except for the natural antioxidant compounds, sometimes additives, such as vitamin C are added to beer, contributing to the overall antioxidant capacity of the final product (Preedy and Rajendram, 2009).

HOPS AS SOURCE OF BIOANTIOXIDANTS

Hop (*Humulus lupulus* Linnaeus) has been long used as medicinal plant because of the anti-inflammatory effects and anti-microbial properties (Karabin et al., 2016). Since ancient times hop has been used against constipation, foot odor, leprosy and for blood purification (Karabin et al., 2015). The brewing applications of the hops are mentioned from the 13th century. Starting from the 19th century, the phytotherapeutic effects of hop against sleeplessness were explored (Dimpfel and Suter 2008). Moreover, because of the presence of bitter components hops have been found to stimulate the gastric secretion without affecting the acidity (Kurasawa et al., 2005). The hop cones are rich in biologically active compounds of particular interest for the phytopharmaceutical applications (Karabin et al., 2016). Anyway, the hops are nowadays manly used in brewing, because of the α- and β-bitter acids and essential oils found in the hop cones, which are responsible for the bitterness and hoppy aroma of beer.

The α-bitter acids consist mainly of humulone, cohumulone, adhumulone and in small amounts of posthumulone, prehumulon, and adprehumulone (Briggs et al., 2004; Karabin et al., 2016). When boiling the worth at 100°C and atmospheric pressure for 90 min the isomerization of the α-bitter acids occurs to the corresponding strongly bitter tasting iso-α-bitter acids. In a similar manner, the β-bitter acids comprise lupulone, colupulone, adlupulone, prelupulone, and postlupulone which are of minor importance in the brewing process because are insoluble in water, but have many potential biomedical applications, because of their biological activities (Karabin et al., 2016).

When the biosynthesis of α- and β-bitter acids is complete, the lupulin glandes of the hop cones secrete the essential oils, which include several hundred compounds having various sensory, biological and/or physicochemical properties (Briggs et al., 2004). According to Sharpe and Laws (1981), the hop oils can be classified as:

- hydrocarbons (50-80% of the hop oils) including monoterpenes (examples: α- and β-pinene, myrcene, and limonene), sesquiterpenes (examples: α- humulene, β-farnesene, β-caryophyllene, α- and β-selinene, and γ -muurolene), and aliphatic hydrocarbons;
- oxygenated compounds (up to 30% of the hop oils) including terpene alcohols, sesquiterpene alcohols, and other oxygenated compounds. The most studied oxygenated oil fractions found in hops are linalool, geraniol, caryophyllene oxide, and farnesol;
- sulfur-containing compounds (up to 1% of the hop oils) including thioesters, sulfides, and other sulfur compounds, which lack any biological activity (Karabin et al., 2016).

Hop pellets and extracts are nowadays most widely used for hopping beer. The hopping rate used in brewery, is decided considering the content of α-acid. The composition and antioxidant properties of the hop products highly depend on the hop cultivar and conditioning steps. Lermusieau et al. (2001) examined the reducing power of six types of hop pellets (T90) and of two CO_2 hop extracts and observed that all samples exhibit very high intrinsic antioxidant activity. The antioxidant activity, assessed as the inhibition times of linoleic acid oxidation induced in an aqueous solution by a free radical initiator, varied significant between different hop varieties. The best reducing power was registered for Saaz variety, usually used for flavor compounds. Lermusieau et al. (2001) reported no significant participation of the α-acids to the reducing power of the hop. Good correlations between antioxidant activity of hop and total polyphenols and flavonoids contents were found in case of all hop pellets. On the other hand, the hop extracts, which lack polyphenols and have a high level of bitter compounds, provided low reducing power to the boiled wort, which was mainly assigned to the presence of α-acids (Lermusieau et al., 2001).

MALTS AS SOURCE OF BIOANTIOXIDANTS

Cereals are rich in phenolic compounds which act through different mechanisms, providing protection against chronic diseases, including different types of cancer and coronary heart disease. Nardini and Ghiselli (2004) reviewed some of the most important biological effects of the phenolic compounds of vegetal origin: scavenging the free radical, metal chelation, interference with the activity of different enzymes and signal transduction, and activation of transcription factors and gene expression.

Barley is the most often malted cereal grain. The majority (about 80%) of phenolics found in beer arise from barley malt (Goupy et al., 1999). Other cereals subjected to malting are wheat, sorghum, rye, out and millets (Briggs et al., 2004).

The antioxidant activity of barley, the total phenolic content and the profile of the phenolic compounds vary among different cultivars and varieties, and changes significantly during malting (Goupy et al., 1999; Lu et al., 2007; Zhao et al., 2008; Sharma and Gujral, 2010). For instance, Sharma and Gujral (2010) reported total phenolic contents ranging from 3.07 to 4.48 mg ferulic acid equivalents (FAE)/g, whereas Madhujith and Shahidi (2006) found significantly lower TPC ranging from 0.81 to 1.38 mg FAE/g. In particular, barley and its products have a high content of phenolic acids, proanthocyanidins, tannins, flavonols, chalcones, flavones, flavanones and amino phenolic compounds.

Goupy et al. (1999) isolated the phenolic compounds from nine varieties of barley and followed the effect of malting on the antioxidant composition and activity. They suggested that the flavan-3-ols, which are the major class of phenolics in barley, are primarily responsible for converting the lipid radicals into stable products, therefore preventing the occurrence of "cardboard off-flavour" of beer during storage. According to Goupy et al. (1999), the monomeric ((+)-catechin and (-)-epicatechin), dimeric (mainly prodelphinidin B3 and procyanidin B3) and trimeric (procyanidin C2 and proanthocyanidin T1, T2 and T3 trimers consisting of (+)-catechin and (+)-gallocatechin units) flavan-3-ols found in barley, account for 70-82% of the total phenolic content, which is higher than the levels previously reported by McMurrough et al. (1983).

The phenolic acids found in barley are mainly benzoic and cinnamic acid derivatives. In particular, the hydroxybenzoic and hydroxycinnamic acids are available in free forms or are present as glycosidic esters (Shahidi and Naczk, 1995). The hydroxycinnamic and hydroxybenzoic acids exert their antioxidant activity by acting as free radical acceptors and chain breakers (Goupy et al., 1999). Important amounts of ferulic acid were found in the aleurone layer and endosperm of barley, in addition to p-coumaric and vanillic acids (McMurrough and Byrne, 1992). Moreover, the glycosidic esters of these compounds were detected in husk, testa and aleurone cells.

In addition to the polar phenolic compounds, barley ant malt also contains apolar compounds, such as carotenoids and tocopherols which exert antioxidant activity.

Different antioxidant properties and radical scavenging activity levels were reported for various barley cultivars and significant positive correlations were established between the total phenolic content and the antioxidant activity of the whole grains (Beta et al., 2005; Lu et al., 2007; Sharma and Gujral, 2010). Raw barley has antioxidant activity ranging between 17.01% and 24.92% (Sharma and Gujral, 2010), and the 1,1-diphenyl-2-picrylhydrazyl (DPPH) radical scavenging activity was reported to range between 11.90 and 12.56 µmol Trolox equivalents (TE)/g d.w. (Lu et al., 2007), or between 9.33 and 11.78 56 µmol TE/g d.w. (Zhao et al., 2008). Leitao et al. (2012) assigned to catechin, procyanidin B3 and prodelphinidin B3 the largest contribution (53%) to the total antioxidant activity of barley.

EFFECT OF MALTING ON ANTIOXIDANT ACTIVITY

Malting is initiated through steeping, which ensures grain hydration, required for germination. Physical modification of endosperm, which occurs during germination, is accompanied by the increase of the levels of hydrolytic enzymes needed during mashing and of bioactive compounds. Anyway, the total phenolic content of different barley cultivars was reported to significantly decrease (by 22.24-38.77%) during the first 12 hours of germination, followed by the slight increase (by 0.18-6.49%) in the next 12 hours (Sharma and Gujral, 2010). Other studies indicated that malting resulted in increased phenolic content and antioxidant activity compared to the unmalted barley, the most significant effects being registered in the later stages of germination and subsequent kilning (Friedrich and Galensa, 2002; Samaras et al., 2005; Lu et al., 2007; Qingming et al., 2010; Leitao et al., 2012).

Overall, Goupy et al. (1999) showed that all phenolic classes are affected by malting. Among flavan-3-ols, the degradation of monomers was more intense than dimers and trimers. No significant impact of malting on the presence of flavan-3-ol (monomers and oligomers) in respect to the total phenolic compounds of barley malt (69-85%) were reported. The procyanidin B3 and prodelphinidin B3 dimers prevailed both in barley and malt (65-72% and 53-83%, respectively). During malting, the level of (+)-catechin relative to the total flavan-3-ols decrease from 13-40% to 5-20%, whereas the (-)-epicatechin remained very low, or even disappeared (Goupy et al., 1999). This trend regarding the decrease of the level of phenolic compounds during malting was later confirmed by Leitao et al. (2012). The decrease of proanthocyanidins and catechins levels was assigned to the glycosylation reactions occurring during malting (Friedrich and Galensa, 2002). The important loss of flavonols of 64-91% was as well observed after malting, owing to the thermal degradation. Regarding the phenolic acids derivatives, different trends were observed for various malted barley varieties: in case of most investigated varieties the content of phenolic acids decreased by 35-78%, but for two varieties the phenolic acids level increased from 1.0 to 4.4 μg FAE/ g dry weight (d.w.) and from 7.2 to 9.2 μg FAE/ g d.w., respectively (Goupy et al., 1999).

Moreover, Leitao et al. (2012) reported that ferulic and sinapic acids contributed to the highest extents to the antioxidant activity of malt extract, the activity of this compounds being three-fold higher compared to the barley extract. The levels of ferulic acid and p-coumaric acid were as well higher after malting. In fact, Leitao et al. (2012) explained the four-fold higher amount of total phenolic content of malt compared to barley, by the higher concentration of p-coumaric acid, ferulic acid, sinapic acids and one additional unidentified compound in malt. As suggested by Woffenden et al. (2002), they partially explained this increase by the release of the bound molecules during kilning, which further allowed better extraction. Indeed, ferulic and p-coumaric acids are known as the most important bound phenolic compounds in barley grains (Nordkvist et al., 1984). In addition, the more friable nature of malt compared to barley should be also factored when discussing the better extraction of the phenolic compounds (Leitao et al., 2012).

The content of carotenoids and tocopherols of barley and malt highly depends on the variety. In all nine barley varieties investigated by Goupy et al. (1999), carotenoids were represented by lutein (about 90% of the total carotenoids content) and zeaxanthin. The carotenoids content of barley varied from 174.2 to 850.1 ng/g d.w., and the effect of malting on the evolution of these compounds was diverse for different varieties. In case of most malt samples the carotenoids content was lower compared to the corresponding barley, but the 26.40–33.67% increase was noticed for three samples. Regarding the tocopherols, each of the Υ-tocopherol and δ-tocopherol represented over 40% of the total tocopherols amount of the barley samples, which ranged between 9.7 and 44.2 ng/g d.w., while α-tocopherol was below 14%. Losses of 12–32% were observed after malting for five varieties, affecting all tocopherol types, whereas for other four varieties the tocopherols content increased by 15.26–91.35% (Goupy et al., 1999).

Although at the end of the germination period the overall decrease of the total phenolic content levels was observed by Sharma and Gujral (2010) for all barley cultivars, the antioxidant activity was found to increase by 34.67-85.02%. These observations comply with the previous report of Goupy et al. (1999). Various explanations have been provided for the continuous increase of the antioxidant activity, in contrast to the evolution of the total phenolic content. Different compounds with antioxidant activity, such as tocopherols, carotenoids and vitamin C, are synthetized during germination (Goupy et al., 1999; Dicko et al., 2005). In addition, the release of the bound phenolic and other compounds, mainly located in the outer layers of the kernel but also in the endosperm, as a consequence of the intense hydrolytic enzymes activity during germination, provides an important contribution to the antioxidant activity increase (Doblado et al., 2007; Sharma and Gujral, 2010).

More mechanistic details which might explaining the increase of the antioxidant activity during malting are provided by Goupy et al. (1999), who tested the antioxidant activity of the pure phenolic compounds isolated from nine varieties of barley and their corresponding malts. They showed that, among all phenolic compound, the flavan-3-ols were the most efficient in exerting the DPPH radical scavenging activity. The spatial configuration of the flavan-3-ols influence the efficiency in exerting the antioxidant activity. In this respect it was demonstrated that (+)-catechin is more efficient compared to its isomer (-)-epicatechin and (-)-epigallocatechin. The antioxidant activity of flavan-3-ols appeared to significantly increase with the number of catechin units. Three different methods employed for quantifying the antioxidant properties of the polar and apolar extract obtained from barley and malts, based on DPPH free radical scavenging, inhibition of lipoxygenase activity and inhibition of β-carotene cooxidation in a linoleate model system, highlighted the great importance of the presence of the -OH group in galloyl derivatives (3',4',5'-hydroxyl group on the B ring) to the antioxidant capacity. Goupy et al. (1999) showed that prodelphinidin B3 (dimer of catechin and gallocatechin), is more efficient than procyanidin B3 (dimer of catechin). These findings contradict the previous theory of Irwin et al. (1981), who suggested that prodelphinidin B3 might act as a prooxidant leading to the formation of reactive oxygen species out of molecular oxygen in the presence of copper ions, unlike the dihydroxy flaavan-3-ols, like catechin and procyanidin B3, which act as antioxidants.

The presence of phenolic hydroxyl groups attached to ring structures was associated to the strong antioxidant activity of flavonoids, acting as scavengers of free radicals (superoxide, peroxyl) by donating hydrogen atoms to lipid radicals. Similarly, the antioxidant activity of the phenolic acids, which are small molecules acting as free radical acceptors and chain breakers, was reported to depend on the number of hydroxyl groups. The hydroxycinnamic acid derivatives such as chlorogenic, vanillic, caffeic, p-coumaric and ferulic acids appeared more efficient than their benzoic acid analogs, such as syringic and sinapic acids (Goupy et al., 1999).

When comparing the antioxidant activity of the tocopherols found in barley and malt, Goupy et al. (1999) concluded that Υ-tocopherol is the most potent inhibitor of lipoxygenase activity, being followed by δ- and α-tocopherol.

The highest amounts of total phenolic compounds are located in the outer layers of the kernels. In particular, the pericarp and aleurone layers are rich in antioxidant compounds. Therefore, the phenolic compounds are not evenly distributed in the milling fractions used in the mashing process. Sharma and Gujral (2010) found 4.18 - 4.69 mg FAE/g bran resulting from milling the germinated barley, which falls within the 4.16 - 6.26 mg FAE/g range, previously reported by Madhujith and Shahidi (2008). These levels are significantly higher compared to the wheat mill streams. Beta et al. (2005) reported TPC of 2.98 – 5.30 mg FAE/g in wheat bran, while Aprodu and Banu (2012) found 3.54 mg FAE/g. The antioxidant activity is as well higher in the bran compared to the whole grain and flour fractions. For instance, the bran fractions have antioxidant activity varying with the barley cultivars in the 34.16% - 57.61% range, which increased to 48.69% - 69.54% after 24 h of germination (Sharma and Gujral, 2010).

IMPACT OF PROCESSING ON THE ANTIOXIDANT ACTIVITY AND PHENOLIC CONTENT OF BEER

Brewing processes have a high impact on the antioxidant activity of beer, mainly because of the reactions involving polyphenols (Pascoe et al.,

2003; Derdelinckx, 2008; Gorjanovic et al., 2010; Leitao et., 2012). Anyway, the individual contribution of different compounds to the antioxidant activity of beer is not actually known, the results available in the literature factoring the complexity matrix (Leitao et., 2012).

Leitao et al. (2011) and Leitao et al. (2012) monitored the total phenolic content and antioxidant activity of the compounds found in wort and beer in different processing stages, such as brewing, boiling, hopping and fermentation. When performing sampling in different stages of beer processing, Leitao et al. (2011) found no important differences in terms of total antioxidant activity between wort extract, boiled hopped wort extract, and beer extract. On the other hand, they reported an important contribution of the hopping and fermentation steps to the three-fold increase of the polyphenolic content. The high content of polyphenols found in beer after fermentation was explained the presence of ethanol which facilitated the better extraction. The following polyphenols with antioxidant activity were identified though chromatographic analysis: protocatechuic acid, catechin, chlorogenic acid, caffeic acid, epicatechin, ferulic acid and sinapic acid. In addition, Leitao et al. (2011) detected three unknown antioxidant compounds more efficient than caffeic acid and epicatechin, which provided the majority (about 64%) of the total antioxidant activity of the beer. In a later study, Leitao et al. (2012) followed the fate, during brewing, of individual phenolic compounds obtained after malting process, while performing heating of the ethyl acetate extract obtained from malt. Under these experimental conditions, brewing resulted in the decrease of the total phenolic content and associated antioxidant activity. Brewing caused the decrease of the concentrations of most phenolic compounds. Only two exceptions were reported: the level of p-hydroxybenzoic acid was not affected and a four-fold increase of sinapic acid was noticed. On the other hand, further processing through boiling resulted in no significant change of the phenolic content, whereas the total antioxidant activity was reported to decrease by 30%. Finally, no important differences in terms of total phenolic compounds and antioxidant activity were observed when comparing the brewed and boiled malt extract samples in the absence and in the presence of the hop extract (Leitao et al., 2012). According to Leitao et al. (2012), the hop extract obtained through CO_2 supercritical extraction did not show any polyphenol content.

Lermusieau et al. (2001) showed that the reducing power of the wort can be improved by the appropriate chose of hop. This aspect is important because at the end of the wort boiling the linoleic acid oxidation leads to trans-2-nonenal formation which at levels of 0.035 ppb imparts to the final products the cardboard off-flavour (Lermusieau et al., 2001). Once produced, the trans-2-nonenal is able to bind to different nitrogenous compounds, which offer protection against reducing activity of yeast during fermentation. Further on, when the pH of the beer decreases, or the storage temperature is inadequate, the trans-2-nonenal is released, affecting the sensory properties of the beer. One way to overcome this inconvenient relies on inhibiting lipid auto-oxidation reactions in the brewhouse, hopefully by means of the natural reducing compounds from raw materials (Lermusieau et al., 2001). A less attractive alternative proposed by Lermusieau et al. (1999) consists on the addition of sulphites to the kettle. Lermusieau et al. (2001) reported that the use of hop pellet allowed obtaining pitching wort with 29-36% higher reducing power. When comparing the wort samples hopped at the beginning or 7 min before the end of the boiling step no important differences were found, suggesting that the reducing compounds from hop are highly soluble at 100°C. On the other hand, when using different pellet varieties for hopping the wort during boiling step, precipitation of the wort polyphenols originating from malt was observed. In the absence of hops with appropriate antioxidant activity, this loss cannot be balanced by the reducing compounds (melanoidins) derived from Maillard reactions (Pflugfelder, 1992, quoted by Lermusieau et al., 2001). In addition to the melanoidins produced during malt kilning, further melanoidins-producing reactions take place during wort boiling. Since the polyphenols content of hop is mainly responsible for the reducing power, Lermusieau et al. (2001) explained that addition of CO_2 hop extract resulted in no significant effect of reducing power of the wort, which is in agreement with the findings of Leitao et al. (2012).

Pascoe et al. (2003) monitored the levels of phenolic compounds and antioxidant activity in different stages of ale brewing. A first decrease of the antioxidant activity was observed after malt milling, and was assigned to the loss of Maillard reaction products formed during milling as a consequence of exposure to the air. This hypothesis is supported by the fact that no changes of the phenolics level was noticed after malt milling. In agreement with previous studies performed on lagers, they observed that different processing steps like mashing, boiling, fermentation, chill-lagering and pasteurization resulted in enhanced antioxidant activity. The increase of the antioxidant activity observed during mashing was assigned to the higher levels of phenolic compounds release from their bound forms as a result of enzyme action. Particular increase of the ferulic and p-coumaric acids was noticed. Because of the advanced extraction of ferulic acid from the aleurone cells of malt and of p-coumaric acid the husk-containing fraction, the antioxidant activity increased during wort separation at 75-80°C through the husk filter bed. Moreover, Moll et al. (1981) suggested that, catechin oligomers are hydrolyzed during wort separation, releasing the monomeric forms, which have better solubility and exhibit increased antioxidant activity. Boiling caused the significant increase of the p-coumaric acid, but since the level of other phenolic compounds was unaffected, Pascoe et al. (2003) explained the increase of the antioxidant activity by 6% through the formation of the Maillard reaction products at elevated temperatures. Further Whirlpool separation resulted in small changes of antioxidant activity. The level of catechin decreased most probably because of the removal of precipitated protein-polyphenol complexes and polyphenol polymers (Pascoe et al., 2003). Although the level of phenolic compound remained constant during fermentation, the antioxidant activity increased because of the metabolic activity of yeast which converted glucose to pyruvate through glycolysis, producing in the same time reduced nicotinamide adenine dinucleotide (NADH). Saha et al. (1988), reported that the maximum level of NADH is reached after 120 hours of fermentation. Another yeast natural metabolite with known antioxidant activity is sulfite, whose level increases in the first 100 hours of fermentation and remains constant afterwards (Dufour, 1991, quoted by Pascoe et al., 2003). The 14% increase of the 2,2'-azinobis (3-ethylbenzothiazoline-6-sulfonic acid) diammonium salt radical (ABTS·+) scavenging activity observed after fermentation was counter acted during the warm rest, when the antioxidant activity decreased by 9% because of NADH conversion into nicotinamide adenine dinucleotide (NAD+) and diminishing sulfite levels. Anyway, the highest antioxidant activity during brewing was noticed after chil-lagering (11.2 µM catechin equivalents) when a 22% increase was registered in respect to the previous processing step. Beer filtration resulted in the most significant decrease (by 40%) of the antioxidant activity. Anyway, Pascoe et al. (2003) reported that the final product was characterized by significantly higher antioxidant activity compared to the original malt. Moreover, in addition to the antioxidant activity increase, higher levels of polyphenols were determined after wort separation and carbonation.

Zhao et al. (2010) investigated the phenolic profile of commercial beers available on Chinese market, and indicated important variation among different brands. In the attempt to clarify the contribution of different phenolic compounds to the total antioxidant activity of beer, they employed different antioxidant activity evaluation methods. Phenolic compounds exert the antioxidant activity through different mechanisms: by acting as reducing agents, by scavenging the free radicals, by quenching the singlet oxygen or by complexing the potential prooxidants (Leitao et al., 2011).

ANTIOXIDANT COMPOUNDS FOUND IN BEER

Oxidation reactions during brewing have a great contribution to the instability of the beer flavor. Various attempts have been made to limit the oxygen pick-up during brewing and packaging, but the oxidative staling of beer is difficult to control. The staling related transformations of the main precursor found in beer is ideally counteracted by the existence of endogenous antioxidants, such as phenolic compounds, which allow delaying or preventing the occurrence of the oxidation processes in beer (Zhao et al., 2010). The antioxidant compounds limit the formation of trans-2-nonenal in beer, by inhibiting the activity of lipoxygenase, which is mainly responsible for initiating fatty acid oxidation during wort production, and the non-enzymatic lipid peroxidation processes (Goupy et al., 1999).

The main classes of phenolic compounds found in beer are phenolic acids, flavonoids, proanthocyanidins, tannins, and amino phenolic compounds (Montanari et al., 1999; Gorinstein, et al., 2000; Zhao et al., 2010). Anyway, not all phenolic compounds possess the same antioxidant activity and some even lack this ability. The compounds with flavonoid structure, such as (+)-catechin were reported to exhibit higher antioxidant activity compared to the non-flavonoid compounds, like phenolic acids, stilbenes, lignans and coumarins. The antioxidant and free radical scavenging activities exerted by flavonoids are influenced by their molecular structure: the existence of double bonds and the number and positions of hydroxyl groups are particularly important. Rice-Evans et al. (1996) showed that flavonoids with 3',4' o-dihydroxy moieties and the 3- and 5-OH groups with 4-oxo function in A and C rings are more active in terms of antioxidant activity than non-flavonoid compounds.

Leitao et al. (2011) showed that gallic acid has the highest antioxidant activity, being followed by epicatechin, caffeic acid, catechin, sinapic acid, chlorogenic acid, protocatechuic acid and ferulic acid. Similar observations were previously reported by Kim et al. (2002). On the other hand, the p-hydroxybenzoic, vanillic, p-, m-, and o-coumaric acids displayed no antioxidant activity.

Zhao et al. (2010) showed that gallic and ferulic acids prevailed in all 34 commercial beer samples, accounting for more than half of the total phenolic content. They also reported high levels of (+)-catechin, vanillic and p-coumaric acids, and significantly lower values for (-)-epicatechin and syringic acid. Nardini and Ghiselli (2004) showed that only some of the phenolic acids found in beer are present as free forms, whereas most of them are present as bound forms. The alkaline hydrolysis allowed releasing important amounts of 4-hydroxyphenylacetic acid, vanillic acid, caffeic acid, syringic acid, *p*-coumaric acid, ferulic acid and sinapic acid. Anyway, the conjugated forms of phenolic compounds appear not to be bioavailable in humans (Nardini and Ghiselli, 2004).

Different levels of total phenolic compounds were reported for different types of beer. Dabina-Bicka et al. (2005) showed that dark beers have higher levels of phenolics (ranging from 520 to 864 mg gallic acid equivalents

(GAE) L-1) compared to the light beer (ranging from 301 to 460mg GAE L-1). Similar results were reported by Zhao et al. (2010) and Piazzon et al. (2010).

Different methods relying on diverse reaction mechanisms are usually employed to assess the antioxidant activity of beer sample. Large variation of the mechanisms behind exerting the antioxidant effects was observed when testing beer samples, and was mainly assigned to the complex composition of the investigated matrix. Samples rich in protocatechuic and caffeic acids exhibited high DPPH radical scavenging activities, suggesting that these compounds are prone to donate hydrogen to free radicals (Zhao et al., 2010). The importance of the compounds with DPPH radical scavenging activities is related to their potential involvement in inhibiting the propagation phase of lipid peroxidation, through the formation of non-radical species, while donating protons to the lipid peroxides or hydroperoxide radicals, the major propagators of the chain autoxidation of lipids (Bamforth et al., 1993).

In addition to the biological value, the presence in beer of the compounds which exhibit DPPH radical scavenging activity is important, because they concur to flavour stability (Zhao et al., 2010). The formation of trans-2-nonenal and other saturated and unsaturated aldehydes, due to lipid oxidation, is one of the main causes of beer staling (Vanderhaegen et al., 2006). In addition to protocatechin acid and caffeic acid, the beer samples rich in ABTS·+ scavenging activity, which also reflects hydrogen-donating ability and might stabilize the active oxygen radicals. Therefore, Zhao et al. (2010) postulated that beer with a high ABTS·+ scavenging activity has good flavour stability. Anyway, the ABTS test fails to show the inhibition of the oxidative process, but quantifies the capability of the tested samples to react with ABTS·+. One major limitation of the test derives from the poor reaction selectivity of ABTS·+ for the H atom donor (Roginsky and Lissi, 2005). Unlike DPPH·, ABTS·+, is able to react with flavonoids, having no –OH group in the B-ring and with aromatic acids with one –OH group (Roginsky and Lissi, 2005).

The superoxide anion scavenging activity of beer is as well important, because superoxide anion yields free radicals like peroxyl, alkoxyl, hydroxyl, and nitric oxide through a Fenton reaction and/or lipid oxidation or nitric oxidation (Ambrosio and Flaherty, 1992; Zhao et al., 2010).

Zhao et al. (2010) suggested that the raw materials used for beer manufacturing and brewing process might have notable influence on the metal chelating activity. The existence of trace quantities of iron or copper in beer ensures oxidation through the conversion of molecular oxygen into reactive oxygen species, which results in the occurrence of off-flavour in beer (Bamforth et al., 1993). Zhao et al. (2010) acknowledged the presence of rather weak-chelating phenolic compounds in the investigated samples, which complies with Miranda et al. (2000), who showed that flavanones and prenylated chalcones, such as xanthohumol and desmethyl-xanthohumol, and nonprenylated chalcones, present in beer and hops, lack the ability of chelating copper ions in vitro. In particular, the syringic acid and (-)-epicatechin lack the ability to chelate metals.

It was suggested that (+)-catechin, (-)-epicatechin, ferulic, syringic, caffeic and protocatechuic acids make particular contribution to the overall antioxidant activity of beer (Zhao et al., 2010) and the changes in the level of (+)-catechin and ferulic acid during brewing resulted in similar trend of the antioxidant activity (Pascoe et al., 2003).

It is therefore generally accepted that the spectrum of natural compounds with antioxidant activity found in beer is wide and largely depend on the composition of the raw materials used for brewing.

The total polyphenols intake of the individuals which regularly consume coffee, tea, wine and beer is mainly ensured by these beverages, which are major sources of phenolic compounds, out of which the esters, glycosides and bound complexes are more often present, while the bioavailable free forms exist in lower amounts. Different research groups investigated the relationships between beer consumption and human health (Preedy, 2009), between the alcohol content of beer and polyphenols absorption (Bourne et al., 2000), and between the amount of polyphenols and the antioxidant activity of beer (Gorinstein et al., 2007; Gorjanovic et al., 2010).

REFERENCES

Agu R.C. 2002. A comparison of maize, sorgum and barley as brewing adjuncts. *Journal of Institute of Brewing* 108:19-22.

Ambrosio G., and Flaherty J.T. 1992. Effects of the superoxide radical scavenger superoxide dismutase, and of the hydroxyl radical scavenger mannitol, on reperfusion injury in isolated rabbit hearts. *Cardiovascular Drugs and Therapy* 6: 623–632.

Aprodu I., and Banu, I. 2012. Antioxidant properties of wheat mill streams. *Journal of cereal science* 56:189-195.

Bamforth C.W., Muller R.E., and Walker M.D. 1993. Oxygen and oxygen radicals in malting and brewing: A review. *Journal of the American Society of Brewing Chemists* 53:79–88.

Beta T., Nam S., Dexter J.E., and Sapirstein H.D. 2005. Phenolic content and antioxidant activity of pearled wheat and roller-milled fractions. *Cereal chemistry* 82: 390-393.

Briggs D.E., Boulton C.A., Brookes P.A., and Stevens R.C. 2004. *Brewing science and practice*. London, CRC Press.

Buiatti, S. 2009. Beer composition: an overview. In *Beer in health and disease prevention* (pp. 213-225). Academic Press.

Dabina-Bicka I., Karklina D., Kruma Z., and Dimins F. 2013. Bioactive compounds in latvian beer. *Rural Sustainability Research* 30:35-42.

Derdelinckx, G. 2008. Polyphenols in wort and beer: state of the art in 2008: where and why? *Cerevisia*, 33, 174-187.

Dicko M.H., Gruppen H., Traoré A.S., van Berkel W.J., and Voragen A.G. 2005. Evaluation of the effect of germination on phenolic compounds and antioxidant activities in sorghum varieties. *Journal of Agricultural and Food Chemistry* 53:2581-2588.

Doblado R., Frías J., and Vidal-Valverde C. 2007. Changes in vitamin C content and antioxidant capacity of raw and germinated cowpea (*Vigna sinensis* var. carilla) seeds induced by high pressure treatment. *Food Chemistry* 101: 918-923.

Dufour J.P. 1991. Influence of industrial brewing and fermentation working conditions on beer SO_2 level and flavour stability. *Proceedings of the European Brewery Convention* 23:209-212.

Friedrich W., and Galensa R. 2002. Identification of a new flavanol glucoside from barley (*Hordeum vulgare* L.) and malt. *European Food Research and Technology* 214:388-393.

Gorinstein S., Caspi A., Zemser M., and Trakhtenberg S. 2000. Comparative contents of some phenolics in beer, red and white wines. *Nutrition Research* 20:131-139.

Gorjanovic S.Z., Novarkovic M., Potkonjak N.I., Leskosek-Chkalovic I., and Suznjevic D.Z. 2010. Application of a novel antioxidative assay in beer analysis abd brewing process monitoring. *Journal of Agricultural and Food Chemistry* 58:744-751.

Goupy P., Hugues M., Boivin P., and Amiot M.J. 1999. Antioxidant composition and activity of barley (Hordeum vulgare) and malt extracts and of isolated phenolic compounds. *Journal of the Science of Food and Agriculture* 79:1625-1634.

Irwin A.J., Barker R.L., and Pipasts P. 1991. The role of copper, oxygen, and polyphenols in beer flavor instability. *Journal of the American Society of Brewing Chemists* 49:140-149.

Karabin M., Hudcova T., Jelinek L., and Dostalek P. 2015. Biotransformations and biological activities of hop flavonoids. *Biotechnology Advances* 33:1063-1090.

Karabín M., Hudcová T., Jelínek L., and Dostálek P. 2016. Biologically active compounds from hops and prospects for their use. *Comprehensive Reviews in Food Science and Food Safety* 15:542-567.

Kim D.O., Lee K.W., Lee H.J., and Lee, C.Y. 2002. Vitamin C equivalent antioxidant capacity (VCEAC) of phenolic phytochemicals. *Journal of Agricultural and food chemistry* 50:3713-3717.

Kurasawa T., Chikaraishi Y., Naito A., Toyoda Y., and Notsu Y. 2005. Effect of Humulus lupulus on gastric secretion in a rat pylorus-ligated model. *Biological and Pharmaceutical Bulletin* 28:353–357.

Leitao C., Marchioni E., Bergaentzle M., Zhao M., Didierjean L., Taidi B., and Ennahar, S. 2011. Effects of processing steps on the phenolic content and antioxidant activity of beer. *Journal of agricultural and food chemistry* 59:1249-1255.

Leitao C., Marchioni E., Bergaentzlé M., Zhao M., Didierjean L., Miesch L., Holder E., Miesh M., and Ennahar S. 2012. Fate of polyphenols and antioxidant activity of barley throughout malting and brewing. *Journal of cereal science* 55:318-322.

Lermusieau G., Liégeois C., and Collin S. 2001. Reducing power of hop cultivars and beer ageing. *Food chemistry* 72:413-418.

Liu Q., and Yao H. 2007. Antioxidant activities of barley seeds extracts. *Food Chemistry* 102:732-737.

Lu J., Zhao H., Chen J., Fan W., Dong J., Kong W., et al. 200). Evolution of phenolic compounds and antioxidant activity during malting. *Journal of Agricultural and Food Chemistry* 55:10994–11001.

Madhujith T., Izydorczyk M., and Shahidi, F. 2006. Antioxidant properties of pearled barley fractions. *Journal of Agricultural and Food Chemistry* 54:3283-3289.

McMurrough I., and Byrne J.R. 1992. HPLC analysis of bittering substance phenolic compounds and various compounds of alcoholic beverages. In *Food Analysis by HPLC*, Ed by Noblet LML, Marcel Dekker, New York, pp 579-641.

McMurrough I., Loughrey M.J., and Hermigan G,P. 1983 Content of (+)-catechin and procyanidins in barley and malt grain. *Journal of the Science of Food and Agriculture* 34:62-72.

Miranda C.L., Stevens J.F., Ivanov V., McCall M., Frei B., and Deinzer M.L. 2000. Antioxidant and prooxidant actions of prenylated and nonprenylated chalcones and flavanones in vitro. *Journal of Agricultural and Food Chemistry* 48:3876–3884.

Moll M., Flayeux R., and Muller P. 1981. Relationship between beer flavour and the compostion of barley and malt. *Technical Quarterly-Master Brewers Association of America* 18:31-35.

Montanari L., Perretti G., Natella F., Guidi A. and Fantozzi P. 1999. Organic and phenolic acids in beer. *LWT-Food Science and Technology* 32:535-539.

Nardini M., and Ghiselli A. 2004. Determination of free and bound phenolic acid in beer. *Food Chemistry* 84:137–143.

Pascoe H.M., Ames J.M., Chandra S. 2003. Critical stages of the brewing process for changes in antioxidant activity and levels of phenolic compounds in ale. *Journal of the American Society of Brewing Chemists* 61:203–209.

Pflugfelder R.L. 1992. Wort reducing power - sources, methods of analysis and influence on beer quality. In *Proceedings of the 5th J. De Clerck Chair: Leuven*.

Piazzon A., Forte M., Nardini M. 2010. Characterization of phenolics content and antioxidant activity of different beer types. *Journal of Agriculture and Food Chemistry* 58:10677–10683.

Preedy V.P., and Rajendram R. 2009. Ethanol in beer: Production, absorption and metabolism. In V.P. Preedy (Ed.), *Beer in health and disease prevention* (pp. 429–440). Burlington, MA; San Diego, CA; London: Elsevier/Academic Press.

Qingming Y., Xianhui P., Weibao K., Hong Y., Yidan S., Li Z., Yanan Z., Yuling Y., Lan D., and Guoan L. 2010. Antioxidant activities of malt extract from barley (*Hordeum vulgare* L.) toward various oxidative stress *in vitro* and *in vivo*. *Food Chemistry* 118:84-89.

Rice-Evans C.A., Miller N.J., and Paganga, G. 1996. Structure-antioxidant activity relationships of flavonoids and phenolic acids. *Free Radical Biology and Medicine* 20:933–956.

Roginsky V. and Lissi E.A. 2005. Review of methods to determine chain-breaking antioxidant activity in food. *Food chemistry* 92:235-254.

Saha, R.B. 1988. Intracellular nicotinamide adenine dinucleotide content of brewer's yeast during different stages of fermentation. *Journal of the American Society of Brewing Chemists* 46:72-76,

Samaras T.S., Camburn P.A., Chandra S.X., Gordon M.H., and Ames J.M. 2005. Antioxidant properties of kilned and roasted malts. *Journal of Agricultural and Food Chemistry* 53:8068-8074.

Shahidi F., and Naczk M. 1995. *Food Phenolics: Sources, Chemistry, Effects and Applications.* Technomic Publishing Company, p 32.

Sharma P., and Gujral H.S. 2010. Antioxidant and polyphenol oxidase activity of germinated barley and its milling fractions. *Food Chemistry* 120:673-678.

Sharpe F.R., and Laws D.R.J. 1981. The essential oil of hops—a review. *Journal of the Institute of Brewing* 87:96–107.

Tafulo P.A.R., Queirós R.B., Delerue-Matos C.M., and Sales, M.G.F. 2010. Control and comparison of the antioxidant capacity of beers. *Food research international* 43:1702-1709.

Vanderhaegen B., Neven H., Verachtert H., and Derdelinckx G. 2006. The chemistry of beer aging – A critical review. *Food Chemistry* 95:357–381.

Woffenden H.M., Ames J.M., Chandra S., Anese M., and Nicoli M.C. 2002. Effect of kilning on the antioxidant and pro-oxidant activities of pale malts. *Journal of Agricultural and Food Chemistry* 50:4925-4933.

Zhao H., Chen W., Lu J., and Zha, M., 2010. Phenolic profiles and antioxidant activities of commercial beers. *Food Chemistry* 119:1150-1158.

Zhao H., Fan W., Dong J., Lu J., Chen J., Sha, L., et al. 2008. Evaluation of antioxidant activities and total phenolic contents of typical malting barley varieties. *Food Chemistry* 107:296–304.

Zhou B. and Liu Z.L. 2005. Bioantioxidants: From chemistry to biology. *Pure and applied chemistry* 77:1887-1903.

INDEX

#

2,3-pentanedione, 21

α

α-acids, 10, 127

β

β-glucan, 5, 16, 26, 38

A

alcohol fermentation, 11, 18, 19, 39
alcoholic beverage, ix, 65, 66, 69, 72, 74, 78, 79, 80, 81, 83, 84, 94, 99, 101, 104, 105, 106, 109, 110, 118, 124, 143
aldehydes, 16, 20, 40, 139
amino acids, ix, 11, 15, 16, 17, 18, 20, 22, 25, 42, 48, 124
ancient beer styles, 87
anthocyanides, 24
antioxidant activity, vii, viii, ix, 2, 27, 41, 43, 49, 54, 56, 57, 58, 59, 61, 63, 65, 69, 72, 74, 123, 124, 125, 127, 128, 129, 130, 131, 132, 133, 134, 135, 137, 138, 139, 140, 141, 143, 144
antioxidant potential, 10, 24, 25, 26, 27, 28, 30, 31, 36, 41, 43, 52
aroma, ix, 3, 6, 7, 8, 9, 10, 11, 16, 17, 19, 20, 21, 22, 30, 31, 38, 39, 48, 58, 80, 82, 97, 101, 102, 107, 123, 125, 126

B

barley, 3, 4, 5, 7, 8, 13, 14, 15, 18, 24, 25, 29, 30, 38, 42, 48, 52, 53, 54, 55, 57, 58, 60, 61, 62, 63, 64, 85, 124, 125, 128, 129, 130, 131, 132, 133, 141, 142, 143, 144, 145
beer, v, vii, viii, ix, 1, 2, 3, 4, 5, 6, 7, 8, 9, 10, 11, 12, 15, 17, 18, 19, 20, 21, 22, 23, 24, 25, 26, 27, 28, 29, 30, 32, 36, 37, 38, 39, 40, 41, 43, 52, 54, 55, 56, 57, 60, 61, 62, 63, 64, 65, 67, 70, 72, 75, 77, 78, 79, 80, 81, 82, 83, 84, 85, 86, 87, 88, 89, 90, 91, 92, 93, 94, 95, 96, 97, 98, 99, 100, 101, 102, 103, 104, 105, 106, 107, 108, 109, 110, 111, 112, 113, 114, 115, 116, 118, 120, 121, 123, 124, 125, 126, 127,

128, 129, 133, 134, 135, 137, 138, 139, 140, 141, 142, 143, 144, 145
beer culture, 78, 95, 110
beer quality, 6, 21, 22, 39, 91, 99, 102, 105, 108, 144
bioactivity compounds, 2
biological properties, 26
biomass, 12, 19, 20, 22, 39, 40, 66, 70, 74, 75
boiling, viii, 2, 7, 9, 18, 19, 125, 126, 134, 135, 136
bottom-fermenting yeasts, 11, 12
brewers' spent grains (BSG), 18, 37, 41, 42, 43, 44, 45, 47, 48, 49, 56, 58, 59, 60, 68, 71, 73
brewing, viii, ix, 2, 3, 4, 5, 6, 8, 9, 10, 15, 18, 27, 30, 31, 39, 41, 48, 52, 54, 56, 57, 58, 59, 61, 63, 65, 66, 70, 72, 74, 75, 80, 84, 85, 86, 87, 89, 91, 92, 95, 96, 99, 102, 107, 110, 111, 113, 114, 124, 125, 126, 133, 134, 135, 137, 140, 141, 142, 143, 144, 145
brewing industry, 3, 4, 5, 6, 9, 30, 39, 41, 48, 54, 89

C

caffeic acid, 10, 41, 134, 138, 139
carbonyl compounds, 20, 23
catechin(s), ix, 24, 25, 27, 41, 123, 129, 130, 132, 134, 136, 138, 140, 143
compounds, ix, 10, 15, 20, 23, 24, 25, 26, 27, 32, 36, 39, 48, 123, 124, 125, 126, 127, 128, 129, 130, 131, 132, 133, 134, 135, 136, 137, 138, 139, 140, 141, 142, 143
consumer behavior, 94
consumption experience, viii, 77, 80, 82, 97, 98, 99
corn, 4, 6, 7, 8, 13, 18, 29, 38, 45, 58

craft beer, vii, viii, 77, 78, 79, 80, 81, 82, 83, 84, 85, 86, 87, 88, 89, 90, 91, 92, 93, 94, 95, 96, 97, 98, 99, 100, 101, 102, 103, 104, 105, 106, 107, 108, 109, 110, 112, 113, 114, 115

D

degradation, 23, 25, 30, 45, 46, 52, 124, 130
degree of steeping, 13, 14
design of experiments, 34
diacetyl, 20, 21

E

einkorn, 8
emmer, 8
emotional associations, 80, 100
essential oils, 10, 38, 67, 126, 127
esters, 20, 22, 24, 40, 48, 129, 140
extrusion, 5, 43, 44, 45, 46, 47, 49, 50, 51, 53, 55, 56, 57, 58, 60, 62, 63, 68, 71, 73

F

fermentation, viii, ix, 2, 6, 7, 8, 11, 12, 18, 19, 20, 21, 22, 23, 27, 38, 39, 40, 52, 55, 57, 59, 64, 65, 66, 67, 68, 69, 70, 71, 72, 73, 75, 84, 123, 124, 125, 134, 135, 136, 142, 144
ferulic acid, 24, 25, 27, 36, 41, 51, 55, 56, 63, 128, 129, 131, 133, 134, 136, 138, 140
flavonols, 10, 24, 128, 130
flavour, 63, 64, 78, 120, 125, 129, 135, 139, 140, 142, 143
flocculation, 12

G

gallic acid, 10, 27, 138
germination, 6, 7, 13, 14, 15, 16, 17, 24, 26, 124, 130, 131, 133, 141
glycosides, 24, 48, 140

H

higher alcohols, 20, 21, 40
hops, vii, viii, ix, 2, 9, 10, 18, 24, 28, 38, 110, 123, 124, 125, 126, 127, 135, 140, 142, 145
humolons, 28, 38
hydrolysis, 14, 15, 16, 18, 26, 29, 38, 138
hydrolytic, 14, 16, 130, 132
hydroxycinnamic, 10, 24, 51, 55, 63, 129, 133

I

in vitro, 41, 46, 51, 56, 61, 64, 140, 143, 144
in vivo, 41, 61, 144
industrial beers, 79, 80, 84, 91, 99, 102, 105, 114
iso-α-acids, 10

K

kamut, 8
kilning, 13, 14, 15, 16, 17, 24, 25, 26, 31, 44, 56, 58, 130, 131, 135, 145

L

lautering, 3, 5, 16, 18, 23, 38, 44
lignans, ix, 24, 123, 138
lignin, ix, 24, 42, 47, 123

M

maillard reaction, vii, viii, ix, 2, 17, 24, 25, 32, 41, 43, 46, 47, 53, 64, 123, 135, 136
malt, vii, viii, ix, 2, 3, 4, 5, 6, 7, 8, 9, 10, 13, 14, 15, 16, 17, 18, 23, 24, 25, 26, 28, 29, 30, 31, 32, 33, 34, 35, 36, 38, 39, 41, 42, 43, 44, 48, 51, 52, 53, 54, 55, 56, 58, 59, 60, 61, 63, 64, 65, 72, 85, 89, 113, 123, 124, 125, 128, 129, 130, 131, 133, 134, 135, 136, 142, 143, 144
malting, viii, ix, 2, 3, 5, 6, 8, 9, 24, 29, 31, 38, 42, 52, 53, 55, 57, 58, 61, 124, 125, 128, 130, 131, 132, 134, 141, 143, 145
mash, 5, 6, 7, 16, 18, 26, 36
mashing, viii, ix, 2, 5, 6, 7, 18, 26, 30, 36, 37, 38, 56, 124, 125, 130, 133, 136
maturation, 11, 18, 19, 21, 22, 23
melanoidins, ix, 17, 25, 27, 43, 52, 59, 61, 64, 123, 125, 135
micro-breweries, 80
millet, 4, 6, 60, 63
milling, 18, 36, 63, 133, 136, 145
mixture modeling, 34

N

non-enzymatic browning, 25
nutritional, vii, 2, 26, 27, 29, 46, 50, 51, 61, 78, 81, 124

O

off-trade consumption, 91
on-trade consumption, 91, 106
organic acids, 11, 20, 22

P

phenolic, vii, viii, ix, 2, 23, 24, 25, 26, 27, 28, 30, 31, 36, 37, 39, 42, 43, 48, 49, 54, 55, 56, 57, 58, 59, 60, 63, 64, 123, 124, 125, 128, 129, 130, 131, 132, 133, 134, 135, 137, 138, 140, 141, 142, 143, 144, 145

phenolic compounds, vii, viii, ix, 2, 24, 25, 26, 27, 28, 30, 31, 36, 39, 42, 48, 49, 54, 55, 58, 59, 63, 123, 124, 125, 128, 129, 130, 131, 132, 133, 134, 135, 137, 138, 140, 141, 142, 143, 144

phenolic content, viii, ix, 2, 24, 37, 43, 57, 123, 128, 129, 130, 131, 134, 138, 141, 143, 145

polymerization processes, 25

polyphenols, 10, 23, 25, 26, 28, 37, 43, 48, 57, 61, 62, 70, 127, 133, 134, 135, 137, 140, 141, 142, 143

premium beer, 94, 96

preventive effect, 47

proanthocyanidin(s), ix, 10, 24, 27, 48, 123, 128, 129, 130, 138

product craftsmanship, 78, 97

products, vii, viii, ix, 2, 13, 15, 17, 19, 21, 22, 24, 25, 26, 27, 29, 32, 35, 39, 40, 41, 43, 45, 46, 47, 48, 49, 53, 54, 56, 70, 79, 80, 91, 94, 96, 97, 100, 103, 105, 111, 112, 115, 119, 121, 123, 124, 125, 127, 128, 135, 136

Q

quality, 1, 10, 12, 13, 14, 16, 17, 18, 22, 30, 43, 48, 50, 68, 69, 70, 73, 74, 75, 78, 79, 80, 82, 92, 94, 97, 98, 99, 101, 104, 105, 106, 107, 109, 110, 111, 112, 113, 114, 116, 118, 119, 120, 124

R

revolution, v, vii, viii, 77, 78, 79, 88, 89, 92, 110, 113

riboflavin, 42, 48, 56

rice, 4, 7, 8, 13, 18, 29, 50, 61, 138, 144

roasting, 17, 24, 25, 26, 43, 48, 53, 59, 64

rye, 4, 7, 9, 128

S

Saccharomyces boulardii, 37
Saccharomyces carlsbergensis, 11
Saccharomyces cerevisiae, 11, 58
Saccharomyces pastorianus, 11
search for authenticity, 100
self-expression, 104, 108
sensory attributes, 79, 80, 81
simplex-centroid, 35
simplex-lattice, 34
sirinic acid, 27
small production, 84
social status, 95, 103, 108
sorghum, 4, 7, 18, 55, 128, 141
spelt, 7, 8
steeping, 5, 6, 7, 13, 14, 15, 26, 130
sucrose, 19, 29, 38
sulfur-containing compounds, 20, 22, 127
sustainability, viii, 2, 41, 71, 141

T

tailor-made concept, vii, viii, 2, 30, 31, 32, 33, 35, 37
tailor-made food, viii, 2
thermal exposure, 45
thiamine, 42, 48
top-fermenting yeasts, 11
traditional ingredients, viii, 77, 85, 86
triticale, 4, 8, 9

Index

V

vicinal diketones, 20, 40
vicinal diketones 2,3-butanedione, 21

W

water, 5, 10, 13, 14, 15, 16, 18, 26, 44, 48, 49, 61, 124, 125, 126
wheat, 3, 4, 8, 9, 13, 29, 30, 36, 38, 42, 45, 48, 49, 51, 52, 53, 54, 71, 85, 128, 133, 141

wort, viii, 2, 5, 6, 7, 8, 10, 11, 12, 18, 19, 20, 21, 22, 26, 30, 31, 32, 36, 37, 38, 39, 42, 52, 53, 63, 65, 66, 69, 72, 73, 74, 125, 128, 134, 135, 136, 137, 141, 144
wort amino acids, 19, 21
wort fermentation, 39, 53

Y

yeasts, ix, 11, 12, 19, 20, 21, 22, 51, 123, 124